LIFE BEFORE

EARTH

LIFE AFTER DEATH

Second Printing June 1976
Printed in the United States of America
World View Press, P.O. Box 15, Fort Lee, New Jersey 07024, U.S.A.

Library of Congress Catalog Card Number:

72-94742

CONTENTS

TABLES

FIGURES

CHAPTER I

The Mysteries of the Being of Man

The origin of man, the earth, the solar system and the universe is hidden in time and in timelessness, before there could have been time and space, before there could have been anything that we know as earth or earthy deposits, or the solid state of matter which forms the foundation of man's knowledge.

Astronomers speculate on how the universe came into existence. The "BIG BANG" theory supposes the galaxies to be the result of a "big bang". What brought such infinitely hugh galaxies together in the first place in order to go "Bang" and spread billions of light years apart, is not touched on or is completely ignored.

Perhaps the reason for ignoring this vital point is that the possibility of the "Big Bang" theory being true is infinitesimally small, probably inversely proportional to the size of its macrocosmic subject.

Quasars and distant galaxies discovered by

means of radio telescopes and other sophisticated means extend man's conception of the size of the starry systems.

Scientists of various specialties periodically propose theories of how the sun, earth and moon were formed and how eventually LIFE started on earth as the result of "natural causes", meaning physical causes.

Astronauts bringing samples from the moon and the unmanned flights serving the same purpose are attempts at unveiling through physical material deposits, the origin of moon, Mars, etc. It is also hoped that the origin of the universe and the origin of life can be solved by bringing back to earth material specimens from as far away as possible. A little calculation will show that the probability of reaching the nearest star system and bringing back a sample within the lifetime of man is nil.

At a speed of 10,000 miles an hour, it will take 70,000 years to travel one light year. The "Apollo" flights averaged much less on the way to and from the moon, and the nearest star is four light years away.

Modern man strives untiringly to unearth relics of his past, seeking clues in archaeological, anthropological and paleontological studies to determine how he and the other creatures which leave skeletal remnants may have appeared in ages past and how he originated.

Discoveries in these fields show that living

creatures existed on earth millions and billions of years ago. Displays of dinosaur bones and those of other extinct species of creatures "billions of years old" are familiar to all. Fossil organisms are estimated to be hundreds of millions of years old and scientists hope that moon rocks five billion years old will be found.

These discoveries have demolished the fundamentalists' concept of the several days of creation. Liberal religious views are therefore forced by incontestable evidence to speak of "eons" as the more probable period for creation to have taken.

Philosophical and theological speculations based on extensive and massive accumulations of skeletal remains in museums all over the world have raised doubts of the basic tenets and beliefs of the religions of the world. Religious beliefs are ascribed to the superstitions of primitive man and to his need for a prop.

Darwin's theory of evolution summed up in the phrase "the survival of the fittest" and extended to include the human race, has been generally accepted. Man is considered to be the end product of natural selective processes. Man, in that view, is considered merely the most highly developed animal. Many men and women live in that belief. Some have become worse than ferocious beasts in their actions and their evil deeds shock even a world that is riddled with crime.

The speculations or theorizing, or in plain words, guessing based on the available evidences, have led to serious questionings about the existence of God. To many of those who hold science to be infallible, and they include drop-outs as well as learned professors, such a slogan as "God is dead" is taken to be fact, proved beyond doubt. Or at least, very likely to be true, for otherwise, how can one account for the facts of science? How deep a hold science has on present day man's thinking is shown by two incidents.

In one, a Russian astronaut, one of the most highly and intensively trained member of a select and elite group among modern man's most respected science, space science, reported that he "did not see God" when he got out of his spacecraft while in space. To his way of thinking, that supposedly proves that God does not exist. In the other case, the citizen of the southern hemisphere who attempted to eliminate the pope did so because "he (the pope) is a symbol of superstition and hypocrisy".

Perhaps the astronaut expected to find Apollo riding his ancient slow horse-drawn chariot and wished to show off his highly computerized multimillion-pound thrust craft.

The combined intelligence and technical ingenuity of hundreds of thousands of the best scientific minds in the world were required to put into orbit a spacecraft weighing a few tons. Yet many of these scientifically trained minds and the world as a whole are apt to think that NO Intelligence is needed to keep planets and stars weighing trillions of tons and galaxies of immeasurable size in unerring orbits.

People of all beliefs have asked: "If there is a God, why does He permit so much killing, crime, hate, wars?"

The inequities found in human life on this planet have also resulted in militant leaders establishing systems of government fostering the denial of the existence of God as is the case in communist countries, to emphasis of social work by some religious groups, and to experimentation with psychedelic drugs by those seeking desperately for direct experience of whatever can be experienced beyond the ordinary.

Disbelief in a higher Intelligence will become more widespread as long as natural science advances and speculations based solely on natural science form the only means in man's attempts to penetrate into the mysteries of the coming into existence of starry systems and also of the formation of earth and all its kingdoms, including man.

Science foresees a leveling process, the eventual cooling and dying of the sun, the erosion of the mountains of earth down to the level of the seas, while the ocean deeps are filled by the erosion of the hills. Finally, so the scientists think, the equilibrium point will be reached, fusion of hydrogen to helium on the sun decreases to such an extent that life on earth ceases. Seemingly, that will give mankind millions of years, up to as long as five billion years to look forward to in which to straighten himself out. It is a long period compared to the five to seven thousand years of recorded history. But, has man even a small fraction of that much time?

In such a system, the existence of man is taken to begin at birth or conception and to end at death. Death of the body is the end of the individual man, as far as science can discover.

This is in direct contradiction to the religions of even the earliest cultures. To ancient man, after death there was Nirvana, Valhalla, the happy hunting ground, paradise, Elysium, Hades, heaven, hell, etc.

Who is right? Ancient man who left monumental, enigmatic structures like the pyramids and Stonehenge whose secrets include precise knowledge of planetary movements as were recently revealed more exactly when computers were called into action, OR modern man who can leave a flaming trail blazing to the moon, Mars, Venus and beyond into orbit around the sun?

The questions arise: Was ancient man as ignorant as scientists think? Or was he far in advance of modern man in some respects? Are there unknown states of existence which are beyond the range of the ordinary senses of man so that he cannot be aware of those states? Are there worlds for which man has not developed the requisite organs of perception? Or could mankind have lost such organs which may have given the founders of ancient religions knowledge of the divine beings of whom they spoke, and of evil, demonic beings also who lived in spheres above, around, and even within man? Tempting him, misleading him, diverting him from cosmic goals?

Are there SUPERsensible worlds? Worlds which may be opened up to man who has developed organs of perception suitable for perceiving them just as the world of light and color is revealed to one born blind when sight is given to him?

THE UNSEEN WORLDS

Radio and TV sets are needed to reveal radio and TV waves which otherwise pass us by unceasingly unnoticed. Cosmic rays have passed by man for millenia unknown to him until recent years.

Are there other secrets hidden from man who has concentrated only on the conquest of the physical world?

Are there UNseen worlds around us besides the solid, liquid and airy worlds which we can see, hear, touch, measure, and weigh? And the electrical world whose existence we "prove" by activating motors and lighting incandescent bulbs; and the magnetic world whose existence is "proved" by the turning of the compass needle or by the magnet's action on iron filings or on a needle?

Are there other worlds interpenetrating the solid, liquid and gaseous ones besides the invisible ones mentioned, but for which no adequate instruments have been devised, or can or ought to be devised in the way mechanistic science employs?

In order to answer these questions, let us consider man as he appears before us. How does a sleeping man differ from a corpse, from man in a dream state, and from man in a fully awake, conscious state?

Let us study these conditions of man one by one, observing them without delimiting ourselves with preconceived ideas of man's capabilities, but by using our more or less developed sense of truth and logical thinking and by employing the most exacting scientific attitude possible.

THE BODY OF FORMATIVE LIFE FORCES OR THE ETHERIC BODY

The corpse is like a dead mineral. The corpse begins to lose its organic structure as soon as life ceases. Then it is subject to the laws of chemistry and physics and falls apart. In shorter or longer periods, it decomposes into its inorganic components.

In sleeping man, on the other hand, there is life: the worn-down organism is rebuilt, rejuvenated. Depending on his age, the human being even grows and the most complicated chemical processes take place within him. His glands produce hormones so complex that our foremost scientists in the finest laboratories can duplicate only a few of them, and do so with the greatest difficulty. Much less are the scientists able to figure out in the first place that such and such a precise structure of molecules is necessary to depress or stimulate this or that organ at the right time, to the right extent and in the right way.

However, those who hold the Darwinistic and scientifically held viewpoints of man's coming into being, make exactly this claim: i.e., that the awesome body of man with its complicated digestive system, nervous system, respiratory system with their complex details in physical structure, biochemical structure and function is the result of blind chance!

The sequence in which the nearly two hundred amino acids of the pituitary growth hormone is put together staggers the imagination. To consider the GH (growth hormone) to be the result of chance or as the result of haphazard mutations is scientific fiction of the highest order. Actually that is gross superstition, as gross as that which modern science indicts the knowledge of the ancients.

If a man dies in his sleep, it is obvious that his physical organs are all still in his body. However, no rebuilding takes place. Only physical breakdown goes on. The body begins to return to the elements of inorganic nature.

What has caused this? What change has taken place?

Medical science would say, "The heart stopped beating".

If that is all there is to it, a good mechanic

should be able to attach a device to the heart and keep it beating after death. But can pacemakers, etc. bring back life?

If the engine of an automobile dies because the battery is totally discharged and no current flows, do we say the engine died because the motor stopped? Or do we go deeper and find the real reason, i.e., the invisible electrical current is used up and is no longer present. The invisible forces are gone, so the engine dies.

Likewise, at the moment of death, the body of invisible formative forces, the life body or call it what you will, such as the etheric body, separates from the physical body and man dies. The heart and all the other organs stop because the etheric heart and all the other etheric organs of the body have withdrawn from their physical counterparts.

That is the difference between the sleeping man and the corpse!

At death, an invisible member of man's being separates from the body. This is the body of formative life forces. It is a real independent entity, a body of highly organized living forces within man which first calls forth the physical materials to life and in whose pattern earthly materials are formed first within the womb, and after birth continues to build the human organism step by marvelous, awesome step from babe to full grown man and woman. Throughout life, this entity unknown to physical science, interpenetrates man's entire physical body with ever-streaming, pulsing, weaving forces which cause all life functions to take place in the physical body and gives each man his unique FORM.

For short, let us call this the ETHERIC BODY, characterized as above. In order to "see" this organism, man requires an awakened "spiritual eye", an organ each man can develop as will be indicated later. A true clairvoyant having fully developed spiritual organs sees the etheric body just as we see color with our physical eyes. (Theosophy 17)*

However, it is not necessary to see the etheric body in order to prove its existence. Clarity of thinking can lead to knowledge of its existence.

"Impossible", says the materialist, "because we cannot see it, weigh it, or measure it by means of any instrument we can devise". "We cannot photograph or detect any body emerging from a dying person", says another.

Are not these statements the same as those of a man born blind who denies the existence of colors and light because he himself has not seen colors and light? Can anyone make a seed, such as an acorn, and have it grow into an oak? Or make a protein containing 50,000 atoms such as are produced in the living human body, or complicated hormones in a pea-sized structure, as the pituitary gland and have these hormones regulate life

*All books are by Rudolf Steiner (1861-1925), unless otherwise indicated.

processes?

The simple-minded atheist, simple because he thinks the process of creation is simple, may say, "The atoms came together by themselves, for in such a way did man come into existence: out of the primeval BOOM, the primeval mists condensed into vapors, then to water, then to solid earth, and, somehow along the way, a simple amino acid was formed, then a cell, a living cell! And this cell's progeny finally developed into man". (Now, where did I miss the point as to how the amino acid became living? Or how was life added to the cell?) And who formed the primeval mists and BOOM? Did the primeval mists form themselves? Hardly.

This is as clever as those who once thought that frogs developed out of mud. "Mutate the mud into frog. Mutate CHON (carbon, hydrogen, oxygen and nitrogen), by heat, pressure and cosmic rays into a cell", say the mutators. Now, what did cosmic rays mutate from?

Scientists who use various means to mutate plants suppose that haphazard chance mutations brought into existence all the hundreds of thousands of plant and animal species. One of them writes "mutations are the very source of genetic variability, and so they are ultimately responsible for the evolution of all present forms of life." (Scientific American January 1971, Pg. 87). That sounds as if no Intelligence is responsible for all of creation!

The idea persists that if a chemist is able to produce an amino acid or any other substance known to be formed only in life processes, that life is created thereby. In order to have life, etheric formative forces must interpenetrate the material, organizing it, forming it and sustaining it. Since scientists know nothing about such forces, it is inconceivable that they will be able to intelligently create such forces to interpenetrate their chemical creations in the right way. That does not mean, however, that certain entities cannot enter and make use of what man brings about on earth.

Formative life forces of plants are transmitted to the seed or by cuttings to the succeeding generations of plants. In the human being, the etheric body is patterned after that of the parents, but the child still has his own individualized etheric body which is unique. The child's features, form, and size are different from the parents' even if similarities to the parents exist far more than to strangers.

These examples which pervade all living beings show that life must come from beings already endowed with formative life forces.

Mineral substances, however, do come from the living. To mention a few, we see that from the living tree comes dead wood, finally coal and petrified wood. Or from the tiny living polyps, calcareous deposits form to as great an extent as the Great Barrier Reef off Australia which is 1260 miles long and up to 90 miles wide. Another example is the chalk cliffs of Dover formed by microscopic organisms.

In spite of the fact that no instrument exists which the clever brain of man can conjure forth to measure or photograph the departure of the invisible body of life forces, nevertheless the living sleeping man compared to the corpse PROVES the existence of the etheric body. It proves this just as exactly and scientifically as a lighted incandescent bulb proves the existence of an electric current, or a tuned radio set or TV set proves the existence of radio or TV waves. It is extremely significant that the sets must be properly constructed in a physical way with all the necessary parts and then be properly tuned in order that reception of invisible radio and TV waves becomes possible.

Likewise, the properly constructed DNA molecule found by biologists to be responsible for genetic changes and arrangements must be present in order that the specific formative life forces corresponding to and working through the particular physical molecule can unite with those of the subject under study in order that the particular physical changes and arrangements can take place to form the new mutant.

In the electrical field, the electrical switch ingeniously and simply brings a small piece of wire into play and the invisible forces are brought into action and a light or motor is turned on or off. The pre-existing invisible electricity is merely brought into use by the switch. The switch surely does

NOT create the electric current.

In the same way, the DNA molecule does not of itself create a new species or mutant, but it makes use of the pre-existing invisible etheric forces. One DNA molecule may serve as the "Channel 9" button, say, and another the "Channel 20" button in the field of etheric formative forces which stream through our physical world.

Anyone who cannot see that the living body is the exact scientific proof of the existence of the invisible, non-material etheric body of man and that the etheric body exists as an invisible member of man's being within the physical body and without which the latter becomes a corpse, cannot be considered scientific in his thinking and cannot be taken seriously. His observations are too inexact to merit consideration. If he denies the existence of the etheric body because he cannot see it, he may as well deny the existence of magnetic forces because he cannot see them.

When the physical body is in certain unfavorable conditions, the etheric body can no longer interpenetrate it. Separation of the two members takes place and the physical body becomes lifeless. Diseased organs or a shattered body, among other conditions, cause separation of the etheric body from the physical body to take place, and we call this death.

Similarly, when a TV set has a damaged part, the TV signals do not enter the set and we say the set is "dead", although the TV station is still broadcasting. The perfect physical instrument is essential for the invisible to reveal itself in TV set as well as in man, although the latter can be greatly damaged or diseased and still survive.

THE ETHERIC WORLD

The solid body of man can exist only because there is a solid earth outside of him. His body derives from the earth and is a part of it. Without the external solid physical world, man would not and cannot become a solid physical human being having a bony skeleton.

Man's watery nature in blood and fluids can exist only because there is a watery world outside of him: his watery world derives from the external fluidic world and is a part of it. Without the external watery world, man would have no blood or fluids in his body.

Man's air world in his breathing can exist only because it is surrounded by an external world of air of which his is a part. Without the atmosphere of air about us, we would not have a breathing system.

In the same way, man's body of formative life forces, or etheric body, can exist only because there is an etheric world all around him and his etheric body is formed out of the surrounding etheric world and is connected with it throughout his life as earth man.

The center of this etheric source is our sun! The seat of man's individualized etheric body is in his glandular system, but the etheric body nevertheless permeates the entire human organism.

The etheric body can also be seen to exist in the living tree when compared to a dead tree or to lumber. It can also be seen in the sleeping animal compared to the lifeless animal.

The living tree, the sleeping animal and sleeping man are therefore the ultimate proofs of the existence and reality of the etheric body and of the etheric worlds of living formative forces which interpenetrate the visible world.

Mechanical means, therefore, are not the criteria for proving the existence of the etheric body of plants, animals and man. The ultimate proofs already exist in the living plant, animal and man.

THE SOUL OR ASTRAL BODY

Man in deep sleep is alive, but he is not conscious of his surroundings. Sleeping man is like a living plant that has two members within: the physical body and etheric body. The physical body makes man, animal, plant and mineral visible to physical eyes. The etheric body gives life, growth from within, and reproduction to man, animal and plant. In the case of crystals, the etheric forces are in the mother-liquor working from outside inward. Remove the mother-liquor and growth stops immediately.

MAN IN HIS WAKING STATE

When man is awake, he is conscious of hunger, thirst, desires, instincts and passions and he has an awareness of sensations which his senses bring from every direction. Sensations do not belong to a being having only physical and etheric body as the plant, "sensitive plants" notwithstanding. One does not attribute sensations to iron filings just because they move in the field of a magnet. Neither does the movement of a sensitive plant denote sensation as an inner experience.

THE SENTIENT SOUL

Consciousness of sensations is what makes a man who is awake different from himself when he is asleep. Therefore, there must be an invisible bearer of consciousness which leaves his head, especially, when he is asleep. This second member of man's being that is invisible is his soul or astral body. The part of the soul that is aware of sense impressions is the lowest part of the soul. Let us call this the sentient soul.

Animals also have a sentient soul in that they have consciousness of hunger, thirst, instincts and sense impressions.

THE EGO

In contrast to animals, man alone of earthly beings has the power of thinking and of speech. He alone of earthly beings stands upright. When man says "I", he points to a still higher member of his being. He points to his eternal EGO, his eternal spirit. This is his fourth member, his third invisible part.

The ego is a Spiritual Being in the highest sense of the word. It is the "image of God" spoken of in mankind's loftiest teachings. In their egos, men all come from the same Divine Source. Each man and woman has a "drop" of the Cosmic Ego which he must strive to individualize and develop in earth lives within a physical body. By means of pain, suffering, joys, sorrows, sickness, health, lessons learned in defeat and success, courageous feats in physical deeds and of the spirit, man develops faculties, talents, gains knowledge and then wisdom. He will eventually in full freedom come to the realization of his own lofty origin. He will then strive consciously to raise himself to ever greater consciousness, wisdom, and perfection in the physical world in order to make them his own and add them to the eternal part of himself, to his eternal, his higher ego.

When a man speaks of himself in ordinary conversation, he is on the whole, unconscious of his higher ego. He refers mainly to his personality, consisting of a physical body capable of certain accomplishments, and which has an appearance that gives rise to feelings of pride, both positive and negative, of varying degrees, and of a mind or intellect trained in various specialties. He or she may be most concerned as to how this body is fed, attired, transported and housed, and whether signs of wealth and intellectual achievement accompany it.

The higher ego remains in the background in ordinary earthly life.

THE INTELLECTUAL SOUL

With the power of the ego, man is able to think and speak. In his thinking, man has the ability to form concepts about his sense impressions. He develops logical thinking and makes laws and is able to recognize the laws of nature. He thereby develops a part of his soul which can be called the intellectual soul. Except for special cases, this stage of mankind's development had its beginning in the Greco-Roman period of civilization. Aristotelian logic is still studied in universities and this fact attests to the pinnacle it reached in the eyes of scholars since his time.

Mankind, having had such an impulse from one of its leading members, slowly strives to make the new faculty more and more its own throughout the world and in all its individual members, in the subsequent centuries and millenia.

In our age, the powers of the intellectual soul have been directed toward satisfying the demands of the senses and the physical body. It has given the world tremendous technical achievements: air-conditioned comfort in home and travel; high speed on land, sea, and air; luxuries in food and clothing; entertainment in radio, tv, movies, and stage of fabulous design and colossal dimensions. The immeasurable amount of human thinking which is a spiritual activity expended daily to produce newspapers, magazines, radio and TV programs to satisfy men's sensual needs or to induce him to buy more to satisfy those needs, staggers the imagination. Most human work is also bent in that direction: to satisfy the needs of the sentient soul.

THE SPIRITUAL SOUL OR CONSCIOUSNESS SOUL

When man progresses in his thinking he can arrive at eternal truths of the good, the true and the beautiful. The thoughts he can have about the paths of the planets are true today as well as yesterday. They do not depend on him personally. If he makes the eternal truths his own, he unites himself with the eternal. This part of his soul can be called the consciousness soul or the spiritual soul. The latter term is more correct as it is a consciousness of the eternal spiritual that is meant. That is, it is a consciousness not merely of the senses, but of the immortal, of the spiritual.

In order to develop this his highest soul member most effectively, he needs a knowledge of the objective spiritual worlds, of the soul world, and of the etheric world, in addition to the physical world.

The spiritual soul age started in the fifteenth century. We are in the first third of this epoch. In this age, mankind should reach adulthood, but like most individuals, mankind is still wallowing in its adolescence. To reach adulthood means that he must rise to the responsibilities and maturity of adulthood.

In order that this might happen, he needs a knowledge of THE NATURE OF MAN. In order to acquire this knowledge, he must study the spiritual scientific researches of one who has advanced far in the direction indicated above, and who has been able to put into words understandable by a thinking human being in full day consciousness the results of his researches. In acquiring this knowledge of himself, the student gains an insight into the nature of the Creative Spiritual Hierarchies and finally reaches a concept of the nature of God.

The eternal truths which each one can make his own unites him with the spiritual worlds and his intellectual soul and spiritual soul expand the more he makes the true and the good his own. The seer sees the various sheaths of man enveloping him as his aura. "The more the I is lord over body and soul, the more definitely organized, the more varied and the more richly coloured is the aura".

*Theosophy 25

Thus we see that the soul has three parts: the sentient soul, the intellectual soul and the consciousness soul. The sentient soul gives the ego a consciousness of his bodily nature in relation to

All references are books by Rudolf Steiner unless otherwise indicated.

the physical sense world; the intellectual soul mediates man's inner world of thinking and feeling with the impressions of the outer world; the consciousness soul brings to man revelations of the divine spiritual world.

The proof of the existence of the four members of man's being is the living, fully conscious, thinking human being. Take away any of the four and he is not man. Without the ego, he does not think or speak. Without the ego and astral body (they go out together in sleep), man is like a plant. If the etheric body also leaves, man dies.

THE HIGHER MEMBERS IN PLANT, ANIMAL AND MAN

Kingdoms	Members in the physical world					Similar human states
	1	2	3		4	
Mineral	physical body					death corpse
Plant	physical body	etheric body formative life forces				deep sleep & coma
Animal	physical body	etheric body	astral body (sentient soul)			dreaming sleep
MAN	physical body	etheric body	astral body: sentient soul	astral body: intellectual soul; consciousness soul or spiritual soul	EGO (eternal spirit of man)	fully awake
	Decomposes, follows laws of chemistry & physics	Grows rebuilds physical body, reproduces	awareness of sense impressions, feelings, instincts	thinking; consciousness of eternal truths	re-embodies in new physical, etheric & astral bodies in progressing earth lives	

Table 1

If man considers himself to be only a physical being, he is dealing only with a corpse. If he believes that feelings and thoughts are due merely to chemical and electrochemical processes taking place within the body, then the death of the physical body will be taken to be the end of man. If a scientist insists that chemical and electrochemical processes produce feelings and thoughts, then there should be feelings and thoughts within the chemists' test tube every time two reacting chemicals are placed into the test tube or in the reacting chambers of chemical plants for every chemical reaction is in the final analysis, also an electrochemical one. That is definitely not the view held by scientists. However, that is exactly the conclusion that must be reached logically and scientifically if human feelings and thoughts are due merely to chemical processes within the body.

A science that recognizes only one-fourth of the nature of man can give only quarter truths, at the best. One that recognizes one-half or three-quarters of man's nature can give at best half-truths and three-quarter truths. In reality, quarter-truths, half-truths, and three-quarter truths are far more dangerous than outright lies. Lies can usually be detected, whereas it takes considerable knowledge to ferret out the true from the untrue in partial truths. And that is the sad state of modern man. He is befuddled in quarter truths, relying wholly as he does on physical science in all its forms.

Consider the mass murders of entire populations in recent history based on the belief that the destruction of the body by guillotine, starvation, Auschwitzes and wars will lead to a solution of earthly problems. Behind these acts is the superstition that the body of man is the sole reality: destroy the body and you annihilate man: no feelings, no consciousness, no ego, nothing exists anymore. LAND, another form of the solid state of matter concept, is all important. All the wars we have been plagued with can be traced back to land: recover this or that occupied land; wage war to occupy more land; colonize to control more land and its resources; defend our land from being taken!

The aggressor nations of the world surely have found they do not need to conquer America in order to benefit from the wealth of America and other countries of the world. Exporting the best and most useful products which their respective genius and skills can invent, design and produce in order to satisfy the needs of other countries have given Japan and West Germany great material wealth for their own people. Other countries concerned with land recovery, protection and expansion, on the other hand, have wasted their wealth in that pursuit, and their own people's demands for consumer goods have been curtailed.

The fallacies must be corrected in mankind's thinking. And this must be done soon, for it is clear to observers in all countries that mankind stands physically and morally on the brink of disaster. The overkill potential of both the U. S. and USSR hangs like a threatening shroud over all mankind.

What actually happens at death? Can knowledge of what occurs to his invisible members after death be gained by modern man? Must he depend only on religious teachings, dogmas, traditions and confessions of faith? Is the statement "No one has ever died and returned to tell us about it" to be taken as the insurmountable obstacle to man's knowledge of his higher nature?

Belief and faith have served mankind through the age of the intellectual soul which extended into the fifteenth century. Up through that epoch of culture, mankind was still in its adolescence. The authority of religions and the rituals still gave man a feeling that there were higher Beings and that there was a Supreme Being.

In this our age of the spiritual soul development, the individual ego asserts itself. Authorities are cast aside. Not only do parents relinquish guidance of babes in diapers or soon after, but it is considered educationally advanced to do so. Kings and emperors, prime ministers and presidents, priests and bishops, even cardinals and the pope are losing their grip of authority and claim of infallibility in the minds of men in every part of the world, and also among their own followers.

The awe-inspiring accomplishments of science seemingly place man on a par with God. This idea cannot be attributed to the truly great scientists because they surely can see the inexhaustible wisdom in all of creation and realize their meager knowledge even in their own special scientific fields. However, those who have a smattering of knowledge of the accomplishments of science or who hold degrees of diversified sorts, and the common man of little real knowledge can jump to conclusions of his own grandeur just because he may have taken a jet trip, has seen, via TV, men in spaceships on interplanetary trips, or reads science-fiction books, sees TV programs and reads comic strips of time-machines which can transport one to any period of history, past as well as future.

These can be the loudest boasters and the most arrogant, but they are unable to do any of the things they boast about and of which they are so proud.

Discoveries of manuscripts such as the Dead Sea Scrolls and the uncovering of ancient cities mentioned in religious documents do not restore beliefs or change the situation. Even if ten thousand eye-witness testimonies of the crucifixion and resurrection of Jesus Christ were discovered, even if engraved in gold, they would be convincingly explained away by some clever intellectual.

Just suppose a scholar some hundreds of years in the future should find the reader's jawbone. Will he have any idea of what your feelings, thoughts and ideals are? Whatever he says will be merely speculations spun out in his own head, if he should use the methods which present-day scholars employ.

Abetted by our culture's hero-worship of titles, be they academic, military, aristocratic, pugilistic or what have you, any such title-holder can explain away any record of the past as aberrations of the mind of inferior cultures of primitive superstitious unenlightened heathens. For how can anyone be as wise as a modern-day Ph.D. or any other title-holder? Surely not that prince-turned-beggar, as Buddha or anyone who wields a hammer as "Jesus, the simple man of Nazareth", thinks the Ph.D., especially if he may have pored over Dead Sea scrolls or Essene literature and then concocts theories of a drug-saturated-sponge-a-la-Romeo-and-Juliet with a twist, in his attempt to explain away and thereby destroy men's belief in the crucifixion, resurrection and ascension of Christ. Simply by speculating, stronger and stronger attempts will be made to eradicate all human beliefs in the Divine.

Is there some way for man to gain direct, exact knowledge of past events when such powerful forces are brought to bear to destroy all religious beliefs?

The critical question becomes: Can modern man develop organs of perception by means of which his supersensible members: etheric body, astral body and ego can be perceived?

The answer is a definite and unequivocal "YES".

CHAPTER II

The Works of Rudolf Steiner

Rudolf Steiner (1861-1925) in his book KNOWLEDGE OF THE HIGHER WORLDS AND ITS ATTAINMENT* states in the very first sentence: "There slumber in every human being faculties of perception by means of which he can acquire for himself a knowledge of higher worlds". The methods he describes start from man's highest faculty, his thinking, and advances to enhanced states of consciousness, with his own ego ever in control of himself.

First he describes fundamental moods of soul needed for progress: veneration and devotion for truth and knowledge leading to humility, tranquillity to distinguish the essential and non-essential, keener observation, a mobility of thinking, refining of one's feelings; a silencing of one's sympathies and antipathies so as to be able to listen to sounds of animate and inanimate objects

*Knowledge Of The Higher Worlds and Its Attainment will be abbreviated to "KOHW" hereafter.

and to let the sounds they produce reveal their inner nature to him.

By means of these preliminary exercises and others described in the 272 page book (1961 edition), one is guided along safe, sane, mature paths for developing higher faculties which are consistent with the constitution of modern man.

One of the preliminary steps consists in "setting aside a small part of his daily life in which to concern himself with something quite different from the objects of his daily occupation . . . His joys and sorrows, his cares, experiences and actions must pass in review before his soul; and he ought to adopt such a position that he may regard all his sundry experiences from a higher point of view "as if detached from himself . . . . as if observing another." KOHW27 The exercises require truthfulness with himself and demand courage, great effort and will-power.

Out of such acts of will are the seeds sown for real self-knowledge. Coupled with this is the fundamental rule in responsible occult training that the student takes three steps in the perfection of his own character for every one step in the pursuit of knowledge of secret truths". KOHW70

On undertaking the development of higher faculties of cognition, the student seeks self-knowledge as a first step. Knowledge of his own weaknesses and defects of character as well as of his more subtly hidden faults is essential. Otherwise the defects act to distort and blur all he sees and hears in the supersensible worlds, thereby deceiving himself and those who hear of his experiences.

The student needs great effort to correct his faults. For instance, determination and courage are needed to transform laziness to useful activity.

The road to supersensible knowledge is not an easy one. Just these conditions of soul outlined above will show that it runs counter to our modern striving for ease and comfort, short cuts to everything, superficiality and permissiveness. Besides, people do not want to change themselves. They are just too, too comfortable to make the effort.

Yet it is a turning point in human evolution in which we live. Man must take it upon himself to strive upwards towards his true nature which is the nature of a spiritual being, his ego being.

If the student carries out the exercises with perseverance, the qualities of tranquillity and self-assurance pervade his whole being. He contemplates the world around him. Silencing extraneous thoughts, he may observe a real seed, for instance, and carry thought after logical thought and compare the living seed with an imitation: i.e., out of the seed, a plant can grow under the influences of light and earth. Nothing can grow from the artificial seed. He feels intensely the invisible something in the seed and the invisible forces work on his thinking and his soul to help build his higher organs of perception.

He advances to contemplating the development of the seed to full-grown plant, then the formation of the seed and the dying of the plant, but with its life forces embedded in the seed. His thoughts and feelings are concentrated deeply and wholly in the life processes of the plant.

Out of such concentrated thoughts and feelings are the forces active which create man's supersensible organs of perception.

Such are the exercises of soul which gradually add new forces to man's etheric body and astral body and form his higher organs of perception just as repeated lifting of weights add strength and muscles to a weight-lifter's body.

Intensive physical effort is required to develop the physical body especially if it is at first deficient in any way. The exercises of willing and thinking and feeling must be carried out with equal determination and perseverance if the soul and spirit organs are to be formed in a healthy way.

If the steps of this training sound somewhat like those of ancient religions, do not be surprised. Mankind had Teachers and leaders initiated into the mysteries of existence by means of long training in centers known collectively as mystery centers or schools of initiation throughout the development of each culture. The original teachers were seers who were clairvoyant, i.e., could see into the supersensible world at certain times and were aware of the spiritual world as a direct experience.

Although each culture became decadent and at its last third lost its pristine wisdom, and retained merely the outer form, the form that is handed down may still be correct. The form that survived may still be able to express the law of the spirit.

The view often expressed by some of our contemporaries that a teaching is untrue because it is merely a repetition of ancient primitive superstitious teachings, is on the same level as that of a scientist or modern critic who says "Newton's law of gravitation cannot be true because prehistoric man already had a similar belief or superstition that taught that it was not safe to hide under a laden cocoanut tree when a storm swayed the trunk. Some other tribe wearing only loincloths already believed that one should not stand under a cliff when ones enemies at the top of the cliff had rocks and boulders. Newton was just repeating primitive superstitions. Poor benighted Newton" , concludes our would-be authority.

Such thinking also concludes that the incarnation of a God cannot take place, even if it did. "So many primitive religions already expected the coming of a god, even up to recent history. Captain Cook was first taken to be a god by trusting Hawaiian natives; so were the Spanish conquistadores by the new world Indians."

Such would-be authorities would say "the resurrection of Christ is merely similar to rites of spring found among most heathen peoples. Therefore," the critic concludes, "resurrection of Christ did not take place. Those who were there were 'programmed' by their ancient lore to believe that it did."

THE DEMAND OF OUR DAY

Man in our day has reached a high degree of consciousness of himself as a separate being. The individuals, however, still think of themselves as in the previous outgrown age as belonging to this or that nation.

Since the last third of the nineteenth century, humanity entered a new cosmic age when the idea of true brotherhood of man in the EGO, in his ETERNAL SPIRITUAL NATURE, should become real to him! Not each for his own nation only, not the formation of more and more new 'nations' as is taking place in a wholesale way in the last two decades, not the further splitting of humanity into smaller or larger groups, not into unions seeking selfish goals, BUT HUMANITY AS A UNITY should become man's proper concern and development.

At this turning point of time, it is essential that exact knowledge of man's higher nature and of higher worlds around him, and of Divine Hierarchical Beings, as well as of the Tempters of humanity, be acquired in an ego-conscious way. This knowledge must become the common property of all men. Otherwise, the old divisive way of split-up humanity vying with each other and hating those of other origins, or those who live across artificially created or natural borders, will lead to more of the conflagrations which have constantly plagued man throughout time.

In order that human freedom may develop, the presence of the higher beings had to be cut off from man's direct perception. This severance has become complete for most human beings in our age, but there are those who still retain remnants of atavistic or naturally occurring clairvoyance which all men had in the distant past and remnants of which remain to this day.

However, it is man's present and future task to gain EXACT clairvoyant faculties by developing his supersensible organs in his higher invisible members in a healthy, ego-conscious way by means of efforts in concentration, meditation, etc., as given in Rudolf Steiner's book KNOWLEDGE OF THE HIGHER WORLDS AND ITS ATTAINMENT.

By means of the healthily-formed organs, he can then see and hear accurately in the higher worlds. His ego is at the same time fully conscious and he enters not a sub-human world only, but he may also, if he is judged to be prepared by his unseen Guide, cross the Threshold into worlds unknown and unsuspected by modern man.

If he is impatient and uses the wrong methods to enter the higher realms and fails to take the necessary precautions about correcting his faults, it gives rise to deformed soul and spiritual organs. These will produce in those worlds the same bad effects as deformed physical organs do in the physical world.

The stages of development lead from preparation to enlightenment to initiation. It is not the intention to give here an outline of these stages of training. A serious living of the exercises coupled with an understanding of why they are necessary and where they lead cannot be abbreviated into a few pages. It is only the intention to indicate here briefly that there is such a training available for the serious inquirer and to point to the published books which give the full details.

To this end, a few indications of what the training includes and a few of the facts of the mysteries of human existence which have been revealed by means of spiritual scientific researches using the higher faculties developed, will be given.

IMAGINATION

The first stage of higher development reached after long persevering training is that of a picture consciousness called Imagination or IMAGINATIVE COGNITION. This is not to be confused with what is ordinarily meant by imagination, meaning fantasy.

At this stage, the student perceives beings and events in the world next to the visible sense world, and which actually borders this sense world of ours. He sees in a way similar to "seeing" in vivid experiences where one knows that they are more than ordinary dreams. Such experiences are widely known among modern man. The world which opens out before the student is always there.

It is only that he has not heretofore developed his supersensible organs that he cannot perceive them, or does so only in fleeting glimpses or semi-consciously in some dreams. It is there just as the world of color is for him who has sight, but not for him who is blind. This world of picture-consciousness is called the Imaginative world.

There desires, passions, feelings, wishes, sensations, blind instincts, fantasies, and lusts take on myriads of forms and shapes. Besides, supersensible beings higher and also lower than man can assume shape and form, some to entice and deceive him, others to communicate with him. In addition, all the soul qualities concentrated in the various animal species have their group souls in the astral world. (underlined KOHW 170-1)

For instance, the quality of cattiness that embraces all cats, the tigerishness that embraces all tigers, the foxiness of all foxes, the rattiness of all rats: the animal instincts which became fixed in the form of each species of the animal kingdom, the group-souls of each of the species are encountered in the astral world. One can also encounter astral forms representing qualities of supersensible beings which never incarnate on earth. There also are mankind's individual and collective hates, passions, etc. (underlined KOHW 170-1)

What a diversity of shapes and forms all these beings can assume! How hideous some of these are can be sensed from descriptions of nightmares, childhood dreams and feverish experiences. Mankind of former ages had an atavistic or natural clairvoyance which permitted him in a dream-like state midway between our day consciousness and sleep state, to be aware of the astral beings. Recall how art and religions of humanity-in-descending to become a fully earth-being, portrayed gods and demons. Weird and frightful forms are commonly seen, as well as beautiful, alluring and enticing ones. Gargoyles and masks still seen in native dances all over the world depict to a certain extent astral forms and beings encountered by man when he entered the astral world.

In the soul or astral world, the forces of sympathy and antipathy are dominant. Each man's own inner nature attracts to himself those beings of a nature similar to his own. Antipathy repels those unlike himself. Freudian psychologists obsessed with sex will only attract beings of like nature.

Self-knowledge therefore, is essential before entering the astral world, because one's "own desires and passions confront him in animal, or more seldom, in human form." KOHW 171
Feelings, wishes, passions, instincts, and even the condition of one's physical organs are spread out before him and take dramatic or animal form in the soul world and can be mistaken for something objective there. Delusion results for the unprepared and unsuspecting.

To him who knows himself in his deeper aspects, the objective astral world is revealed when what belongs to himself is stripped off what he sees.

He who enters the astral world unprepared and uninformed of its nature will therefore bring back unreliable information and fantasy. The dangers of deception are present every moment. Even one's own inner nature may take on shapes so horrible and frightening that they may be unnerving to the beholder. If he is exposed to them before he is ready for the encounter, it can lead to serious consequences.

Into this astral world is the medium, the drugged and the hypnotized cast. Devilish enticements may be encountered which the unprepared soul and weak ego are unable to cope with. Blind gropings may reveal fantastic sights, but of what use are these? They can only lead to illusion and delusion, a delusion of the subject and the experimenter no matter how scientific and high-sounding degrees he attaches to his name or how honored the institution to which he is connected.

Such dabbling is like child's play compared to the devilishly clever seducers who also work in the astral world. The human being has a supersensible "double" which takes the form of his own inner nature. For the sensuous, undeveloped man, the form can be most hideous. Such a "double" of man is depicted in Bulwer-Lytton's ZANONI. This is one description: "It gazes on thee, the dark, mantled, loathsome thing! There, there, there with their devilish mockery and hateful craft, glare on thee those horrid eyes!"

Man's task is to transform the hideous into a beautiful being, but he must first know that it is he himself in his unpurified inner nature that thus confronts him when he forces his way across the threshold from the sense world into the astral world.

Mankind as a whole has been providentially protected from the encounter thrown open by drugs. The power of the ego is diminished or obliterated by LSD, opiates and other drugs. This is opposed to the course of modern man's development which is toward greater and ever greater ego-strength and consciousness.

Man needs a high degree of moral development, self-knowledge and courage far beyond that possessed by the ordinary human being in order to enter the higher worlds safely.

In the sense world, the correctives are present to keep man from much self-delusion, i.e., he may say that fire is cold, but as soon as he touches it, he finds his error. In the next world bordering the physical, the astral world, man's SYMPATHIES attract to him those soul beings similar to his own desires and wishes. These can therefore, keep him completely unaware of the higher beings in that world.

In recent centuries, natural or atavistic clairvoyance receded as a faculty of man in order that he can gain the faculty of clear logical thinking which must come about for the development of natural science. This was and is necessary so that man could become fully earth man, fully incarnated on earth. Drugs throw man backward to a state from which he had evolved.

Man has now to gain freedom as an individual ego distinct from his former state of being just a member of a family, a tribe, or a nation.

If man continued to see higher, super-human beings whom he considered to be heroes, gods and goddesses, he could not become free.

If one were continually in the presence of a mighty, majestic light-giving Creator, freedom of action would not be possible. One would be compelled by the awesomeness of the Divine Presence to follow the will of the Divine Being. One cannot avoid it in that situation.

INSPIRATION

In the second stage of development called Inspiration, the student can also hear spiritual speech or sounds. Just as on the physical plane we can know very little if we only see a person, but can learn immensely more if we can converse with him and he opens his inner world of thoughts, feelings, ideals and aspirations to us, so too when spiritual speech is heard and the beings speak in inspiration, we can know more about what is seen in the world of imagination.

INTUITION

In developing to the third stage of perception

beyond and above our thinking, called Intuition, man must be able to silence himself so much that he is able to experience what another feels and to live in the innermost being of the other. He learns to observe and maintain, above all, a high moral development which will not infringe on the freedom and sanctity of the being observed, nor to use any of the knowledge he has gained for his personal advantage.

He thus enters consciously where the Creative Spiritual Beings are active in the lives of individuals as well as of nations and mankind as a whole. Their intentions are revealed to him.

In this stage, "man reaches the verities of existence. He attains to Truth and Reality in their eternal aspect, not in the transitory earth state or in the astral state. Communion with higher beings is established. Enlightenment which kindles the spiritual light takes place".

Man rises gradually. He does not attain all this in one great leap. There are many steps in developing to each stage. However, one may take years or not be able to reach a significant level of development in this life, while another may find the spiritual world opened to him unannounced.

In normal life, it takes many incarnations on earth to gain these higher faculties of cognition.

SOME RESULTS OF INITIATION

The faculties of imagination, inspiration and intuition developed in the right way give men the possibility of reading in the hidden spiritual script where "all that has ever happened ON EARTH IS TRACED IN THE SPIRITUAL WORLD AND REMAIN THERE FOR ALL TIME", says Rudolf Steiner. This is known as the Akashic Record.

STAGES OF CONSCIOUSNESS

M A N	Other kingdoms
Deep sleep and coma	plant
Dreaming sleep	animal
Fully awake	present day man while awake
Imaginative Cognition	next higher stage for man
Inspirative Cognition	second higher stage for man
Intuitive Cognition	third higher stage for man

Table 2

THE SPIRITUAL SCIENTIFIC RESEARCHES OF RUDOLF STEINER

THE NATURE OF RUDOLF STEINER'S WORKS

The uniqueness of Rudolf Steiner is that his works (books, lectures, etc.) are the result of exact, direct perception of the etheric, astral and spiritual worlds carried out in full ego-consciousness and which he was able to put into human language understandable by modern man. (Note well that it is not a dimmed-down, trance state that our age requires, but knowledge gained by means of <u>enhanced</u> states of consciousness.)

He has "read" from these records known as the AKASHIC RECORDS or AKASHA CHRONICLE and gave the results in his lectures and books. These can be understood by the human intellect if it is mobile and not fixed. Even as the concepts of natural science change continually and the scientist must not remain fixed in his thinking, the ideas we ordinarily hold as earthmen must not remain fixed when we rise to the concepts of man in his higher nature, in his eternal spiritual nature.

Not only does Rudolf Steiner tell us what his researches reveal, but he shows how each one who strives in the exacting way he describes, can develop such supersensible faculties for himself. Rudolf Steiner is the pioneer who brings the new human capacities of exact clairvoyance to investigate the spiritual worlds consciously. These are essential for man to strive to acquire individually if he is to fulfill his high earthly destiny.

He has opened the vast domains of the invisible worlds of man's higher members to spiritual scientific investigation. He tells us what these higher worlds are like and how man lives in them as well as in the sense world during earth life and also before and after earth life. He gave the results of these researches in some 6,000 lectures.

Many of these have been published and hundreds have been translated into English and are published as books and pamphlets.*

*Note: These are available from book stores and mail order sources dedicated to making

Rudolf Steiner's works available in the English speaking world. Some of these sources are listed in the appendix. (A lending library maintained by the Anthroposophical Society in America at 211 Madison Ave., New York City, will mail books anywhere).

Even if we make little headway in developing our own supersensible organs, studying the results of a spiritual researcher helps us to gain an understanding of the higher worlds, and finally reveals to befuddled modern man the purpose of earth life as well. This is the process in the training of men in all the professions and sciences: doctors, lawyers, scientists, educators. They spend many long years studying the accumulated results of the researches of others in their fields. Only after such lengthy study of others' works are they able to conduct significant researches on their own, and many never do. Yet these latter add immeasurably to human welfare.

The study of these books therefore is a necessity for anyone seeking to understand the mysteries of existence, or to investigate these mysteries himself.

Man cannot safely drive a car while under the dimmed-down consciousness due to the influence of alcohol and drugs. He does not explore space without knowing what to expect in space and what his spacecraft can and cannot do. Astronauts are intensively trained to know of the hazards of space and how to cope with them before they venture into inhospitable space.

The difference between our earth and the moon is common knowledge to TV viewers who see the complicated life-support system worn by astronauts. Without these they cannot survive more than a few seconds on the moon.

The difference between our physical world and the astral-spiritual world is much greater than between earth and moon, and it is of a different nature and therefore much more difficult for our physical-world-oriented mind to comprehend.

The would-be investigator of the worlds of soul and spirit must therefore learn beforehand of the nature of those worlds and of their hazards so that he can enter them safely and with beneficial results for himself and his fellowmen. Curiosity seeking and egotistic dabbling may lead to shattering results.

With the capacities indicated above, Rudolf Steiner was able to penetrate into the secrets of man's entire past. He has read in the "memory of the earth" and of the solar cosmos not only what was outwardly visible or was left in undecomposed remnants such as are needed by our materialistic scientists, but he was able to penetrate into what took place in the thinking and feeling behind the deeds of men, because "thoughts and feelings leave their traces too in the eternal world".

In his THEOSOPHY, Rudolf Steiner reveals the nature of man far beyond what is possible in this sketch. He tells of the life of the ego, astral body and etheric body during waking day consciousness, during sleep and also AFTER DEATH AND BEFORE BIRTH!

In his OCCULT SCIENCE AN OUTLINE, he traces the first beginnings of the creation of the solar cosmos eons before there was any possibility for fossilized remnants to be left for present day paleontologists to uncover.

Stupendous? Fantastic? Unbelievable?

"Horrors, horrors, horrors" will be the cry of the academic world. "We found nothing of the kind!"

Of course not. Neither did the ostrich which stuck its head in the sand.

OUR PRESENT CULTURE

The academic world has been absorbed in the last few centuries in providing specialists who have produced the greatest technological civilization. The average man in this country has at his command greater luxuries than kings and emperors of a few decades ago. He has instant light at the press of a switch, water of the right temperature at the turn of a handle, regal transportation at the turn of a key, entertainment surpassing that of sultans at the turn of a button. Buttons also summon a secretary to a president or to a businessman; buttons control a house heater or an atomic power plant that can light up a city; buttons operate a household mixer or a huge computerized steel mill.

The common man lives in greater ease and luxury than even kings dreamt of just fifty years ago.

In all of these, what have we taken care of by means of the tremendous outpouring of human ingenuity, inventiveness, scientific brilliance?

All these have served the physical bodily needs of man! For his comfort, enjoyment, ease, entertainment!

MEMORY

Our ordinary memory takes us back in TIME to the moment before, the day before, the week, month, years BEFORE in TIME. "The cosmic memory of the Akasha Chronicle takes the one who has acquired Imaginative Cognition, back into the past not in time but in SPACE. The scenes he sees are spatial. The past is stretched out in pictures in space, side by side as it were. The initiate enters into a FOURTH DIMENSION of space unknown to science when he investigates the past." (Knowledge of Higher Worlds)

One can picture it as if all the events of earth are imprinted on a huge cosmic film, every moment as it happens. The next instant the earth and sun have moved to space where they have never been before. We know the sun is taking our whole solar system along with it in the Milky Way. In addition, the earth is traveling at a speed of over one and a half million miles per day in its orbit around the sun. The spiritual record of all that ever happened in the solar system since Ancient Saturn when only spiritual warmth existed can be investigated by the modern Initiate.

CHAPTER III

Man's Supersensible Members After Death

Initiate consciousness is able to say the following, among much more of the invisible members of man's being after death. The etheric body of formative life forces no longer needs to supply the forces needed by the physical body. When it is released from this tremendous task, it is able to bring about a panoramic view of the life just ended. Starting from where it was at death, the scenes unfold backwards to birth. The panorama is best experienced in close connection with the physical body just vacated. It is conducive to the best transition if the physical body is not buried or cremated during the three and a half days after death.

To the uninitiated ego that is not aware of what happens after death, the sudden deprivation of the physical body can be quite a shock and requires a period of adjustment. The presence of the physical body is therefore an aid during the transition period.

At the end of this time, an essence of the etheric body is taken up by the ego as the fruit of the last life, and the rest of the etheric formative forces returns to the surrounding etheric cosmos and becomes a part of it. All our thoughts are therefore added to the etheric world around us.

THE SOUL AFTER DEATH

Since the astral body is the bearer of consciousness, consciousness does not cease at death. When the physical body is laid aside, the soul is released from the stupendous load of caring for the physical body in the midst of life's activities.

With the lifting of this load, the soul and ego are much more conscious than during ordinary day consciousness. In fact the enhanced consciousness is so great that the ego subsequently looks back to the moment of death as a brilliant experience. It is the opposite of the experience at birth when the ego is dimmed down to a point where the new-born babe is practically helpless and wholly unconscious of his earthly surroundings.

"After the period of retrospect, the ego is bound to the soul for a fairly long time in the soul world." (Theosophy 70) The first part lasts approximately one-third of the length of the earth life just completed, or said in another way, it is equal to the length of the periods of sleep in the last life. The enhanced consciousness of the soul and ego review all of the experiences of the life just ended, but in an astonishing way. Each one sees how his deeds affected the other, rather than how they affected himself on earth. One feels bliss as the result of his good deeds to others in the earth life just past, but one also feels the pains and agonies one has caused others.

Consider this fact against the hope of evil-doers who by hiring the cleverest lawyers and by the use of legal technicalities think they can forever escape punishment; or those who do escape punishment even for murder because they "may not incriminate themselves". Men's foolishness committed on the highest legal level cannot negate the moral laws which govern the universe.

After death, the ego and soul of man himself bear witness to his own fall to a less perfect condition because of his own misdeeds. The harshest judge is one's higher consciousness which when stripped of earthly considerations sees oneself as one is in reality.

THE SUBJECTIVE AND THE OBJECTIVE FACTS IN EVIL DEEDS

There are two aspects to every evil deed. The first is how the deed affected the one who was sinned against. Take a person who was blinded by another. The one blinded or maimed will not be able to carry on his life tasks in the same way. He may have to change his work drastically. It may even change the life of those who have to care for him. In addition, the lives of his descendants may be changed and be affected adversely. Whole trains of consequences can result, many of which are far beyond the capacity of the evil-doer to correct or rectify. Those aspects are the consequences of his evil deeds which are "forgiven", that is, those consequences of evil deeds are and must be borne by a Cosmic Being whose nature must be LOVE AND SACRIFICE. This Cosmic Being is the One who sacrificed Himself even for those who do not recognize Him, but who condemn, revile and ignore Him. This Being, in our age, is called Christ.

If man's evil deeds kept piling one on top of the other and remained unforgiven, i.e., not borne by a Cosmic Being of Love and Sacrifice, he would become overwhelmed by them and he would soon sink below the human level.

By our evil deeds, our 'unmorality, unkindness, and lack of interest in our fellow-men' we, believers and non-believers as well, continue to 'crucify Christ' even in our day. (Anthroposophic Ethics 72)

Forgiveness affects what evil deeds do to others, not to what evil deeds do to one's own inner Self. A human being who truly forgives another, foregoes taking the compensating deed for himself. The compensatory deed is given to humanity as a loving sacrifice by the person who truly forgives.

The second aspect of an evil deed is its effect upon the evil-doer. The one who blinds another has, by his evil deed, become less perfect than he was before the act. He has created evil and this becomes a part of his soul and can be observed in his "double's" evil-looking physiognomy.

In order to regain his previous state, it is necessary that he himself perform a good deed of equal import. No one can do it for him. Only what he acquires by his own effort becomes permanently his own. Only what he himself does can change the evil-appearing physiognomy to one of beauty in his supersensible "double".

A confession does not transform the evil doer into a saint. A confession is an admission in his consciousness of his error and the admonition is not to repeat the error: "go and sin no more". Those parts of sinful deeds whereby one becomes less perfect cannot be wiped out by being forgiven. This fact must be a shock to many. The evil deed itself has been "written" into the sand, that is, it is written into the earth and the correction must come on earth in the future.

DESIRES OF THE SOUL AFTER DEATH

Desires do not cease when the physical body is laid aside at death because desires reside within the astral body, within the soul. Therefore, all the desires in which the soul finds enjoyment such as longing for a tasty dish, and which can only be satisfied by means of the physical body, i.e., the taking in of food which, while it satisfied the feeling of hunger and thirst, also gives the soul pleasure and enjoyment, will become literally "burning desires" within the soul because there is no physical body and no taking in of food. Take the glutton who while in the body can eat to satisfy his pleasure in eating. The burning desires will be stronger in him and more difficult to overcome than in one who has less compelling desires.

Add this to all the other desires of earthbound man and the "region of burning desires", the "purgatory" of Christianity, the place of cleansing fire, or hell, etc. can be comprehended.

It must be noted that the desires come from within one's own soul, so each human being creates his own hell.

This condition of burning privation lasts until the soul and ego learn not to long anymore for that which can only be satisfied through the body. (Theosophy 72)

The first period after death when the astral sheaths connected with the lowest part of the soul, the sentient soul, are purified takes place in the region extending outward from earth to the orbit of the moon. How one passes through this region varies with the individual as to how the soul was bound to the physical. It will be short if he has clung little to physical life, long if his interests were completely bound up with it and at death many desires, wishes, etc. still live within the soul. Natures with few lusts go through this region without noticing it. Souls entangled with lusts after death demand its purification because only thereby can an imperfection that exists in the soul be purged away.

Generally in life man's wants, desires and passions are subdued, conquered or become reduced simply because of advancing age, so those who die in advanced years are already eliminating in earthly life some of the ties to the bodily nature which it is necessary to eliminate after death in order that the ego may live in its own world: that of the spirit.

A student of the spirit takes positive steps to transform his desires and weaknesses even while living an active life. He thus hastens his own soul and ego development.

In this first period after death, man's moral nature reveals itself as physiognomy. He can approach only those of similar physiognomies.

On earth, the villain can hide his real nature and a "wolf in sheep's clothing" is possible. After death, this is no longer so. "The physiognomy reveals how far a man was in earth life, a good man or a bad man, a wise man or a fool", says Rudolf Steiner.

"Since in this period, man can approach only those who appear like himself, the inability of the immoral to approach those of moral nature will be like so many judgments passed upon him."

This points to the utter futility of those who seek messages from those dead, who while living on earth ignored the Divine, the moral and the good, as to whether they see God after death.

The soul and ego learn to conquer the desires one by one, the fire of unfulfilled desires consuming them, and the soul and ego is stripped of them gradually.

These stripped off astral sheaths of the so-called dead are what mediums contact, the ego and soul having advanced higher in their development. To consider these astral sheaths the true human being would be like taking the empty cocoon for the butterfly, or the stuffed discarded garments of a human being for the real self.

INTENTIONS FOR THE FUTURE

When the ego and soul see the evil effects of his misdeeds in the earthly life just ended and how they cause harm to those around him, strong intentions are created in the soul to right the wrongs he had inflicted on others. Each error makes the soul have as its aim, the opportunity in the future, to compensate for the errors. A person, or more strictly speaking, the ego therefore ties to his future existence those whom he had injured.

A short resumé from Rudolf Steiner's Theosophy (*Theosophy 70's) is as follows:

"In the second region of the soul world, man disengages from the external glitter of life, the swiftly succeeding impressions of the senses.

"The third region relates to his wish nature. The fourth is connected with the feeling of self as based on his body. After death, the body is lacking and man feels emptied out, lost. This continues until it is recognized that the true man does not lie in the physical. The soul learns no longer to feel the corporeality as essential reality. Suicides suffer especially in this region because they leave the physical body in an artificial way while all the feelings, unsatisfied desires and wishes connected with the body remain unchanged.

"The next stage cleanses the enthusiasm of nature insofar as these have a sensuous character, as distinguished from the seeking of the spirit that reveals itself in the events and things of nature.

"Religious observances leading to an enhancement of sensuous welfare and many other sense-oriented activities of man also fall in this category and must be purified here.

"In the sixth region, the part of the soul that thirsts for action not of an egotistical character, but springs nevertheless from the sensuous satisfaction which action affords them, are purified. Idealists who desire action are in this category; as are those of scientific and artistic natures who give themselves to their work because it pleases them.

"In the seventh region, the soul life is freed from the last inclinations to the sensibly physical world: the belief that its life activities must be devoted entirely to the physical world. This belief in reality is materialistic--the chief conception of this our age.

"Where the soul feels itself to be in affinity,

it remains there until the affinity is extinguished. Where none exists, it proceeds on its way unhindered." (Theosophy 70's)

As the soul expands outward beyond the orbit of the moon, she enters the sphere of the planet commonly called Venus. This sphere includes the belt around the earth with the distance to Venus as the radius. In this sphere, among other Spirits work the spirits of illness and death which bring death and illness not in old age, but in the flower of youth or in childhood through epidemics, illnesses of manifold types and so forth.

There, the seer sees with painful sorrow how those souls who on earth spent a life without a conscience, each condemn himself to become a servant of these evil spirits of illness and death who send these destructive forces into our physical world.

In this sphere also are the after effects of the widespread earthly trait of laziness. The human being having this laziness trait sentences himself to bear the yoke of slavery to the spirits of hindrance who hinder work on earth. These evil spirits are under the domination of Ahriman (see section about this being). (Occult Research Into Life Between Death and a New Birth 3)

He who on earth developed moral feelings, a moral mood in dealing with his fellowmen, sympathy and benevolence easily make contact with those with whom he lived on earth and with whom his soul was united. Those lacking these moral qualities find difficulty in finding those with whom they had been united on earth, while in the Venus sphere, known as occult Mercury.

Many lectures of Rudolf Steiner cover this subject. Among them are Theosophy, Occult Science (Chapter 3), The Soul's Progress Through Repeated Earth Lives and Man's Life on Earth And In The Spiritual Worlds.

The soul and ego expand farther into the belt having the earth as center and a radius equal to the distance of Mercury from earth. All the while, the soul is ridding itself of those qualities which tie it to earth. When it reaches the limits of this Mercury sphere, known as occult Venus, the soul has expanded to the limits of the soul sphere. The ego then separates from the soul, taking only an essence of the astral body with it as the fruit of that incarnation.

Thus it can be seen that there are three corpses: the first is the physical corpse discarded at earthly death, the second is the etheric body which separates and expands into the cosmos about three and a half days later, and the third is the astral body discarded by the ego after the period in the 'inner planets' (moon, Mercury and Venus using the occult terms).

THE THREE CORPSES OF MAN

First corpse: physical body at death.
Second 'corpse': etheric body, 3½ days after death.
Third 'corpse': astral body, period through the inner planets: moon, Venus, Mercury.

Table 3

THE SPIRIT WORLD: THE SUN REGION

The ego at last reaches its true home, the spirit world itself, the sun region, living as a spiritual being among Spiritual Beings. Lofty Spiritual Beings work creatively on the sun!

"Atomic fusion, surface temperatures of a million degrees, core temperatures of ten million degrees" say our scientists who have never been there physically and who NEVER will, even in a spaceship.

If, on a dark night, a physicist or astronomer saw on a distant hill, ten billion fire-flies packed closely together and flashing all at once, how many thousand degrees centigrade do you suppose he would calculate that light source to be? In reality, the temperature would be quite close to that where he is standing.

"The World of the Spirit . . . is so unlike the physical that all that is said about it will appear fantastic to him who is only willing to trust his physical senses". (Theosophy 78)

"This world is woven out of the substance of which human thought consists . . . But thought, as it lives in man, is only a shadow picture, a phantom of its true being. As the shadow of an object on the wall is related to the real object which throws this shadow, so is the thought that makes its appearance through a human brain related to the being in the spiritland which corresponds to this thought. Now when the spiritual sense of man is awakened he really perceives this thought-being, just as the eye of the senses perceives the table or the chair."

Since our language is equipped for expressing earthly, material realities of the senses, it is inadequate to give anything more than an indication of the World of Spirit.

Here Spirit Beings are encountered. "First, the spiritual Archetypes of all things and beings which are present in the physical world insofar as it is not endowed with life. These include the Archetypes of the minerals, the physical form of the plants, of the animals and of man. The physical things and beings are copies of these Archetypes" (Theosophy 79). All is continuous mobile activity, in a ceaseless creating . . . For the Archetypes are creative beings. They are the master builders of all that comes into being in the physical world and the soul world. Their form changes rapidly; and in each Archetype lies the possibility of assuming myriads of specialized forms. They as it were let the different shapes well up out of them, and scarcely is one produced than the Archetype sets about pouring forth the next one from itself . . . Often innumerable Archetypes work together in order that this or that being in the soul or physical world may arise". (Theosophy 80)

"The second region contains the archetypes of life. Here the creative Primal Forces produce everything that appears in physical reality as living beings. It is evident here that the life in man is related to the life of all his fellow creatures." (Theosophy 82)

"The third region contains the Archetypes of all soul formations. Feelings, sensations, instincts, passions, etc. are present in a spiritual way as a gentle zephyr or as a stormy spirit tempest, as the case may be in the other worlds. Nothing of self-seeking clings to the soul in this region, but only what is carried out in selfless devotion to his fellowmen and his community. The benefactors of the human race, those who render great services to communities have gained their ability to render them in this region after having lived an earthlife giving them such a relationship to these Archetypes.

"The Archetypes of the fourth region contain the Beings who govern the Archetypes of the three lower regions, ordering and grouping their activities. What man during his earthlife develops in the way of scientific discoveries, of artistic ideas and forms, of technical conceptions, bears fruit in the fourth region. It is out of this region that artists, scientists, great inventors draw their impulses. For all men, what occupies them beyond the sphere of daily living, wishing and willing originate here.

"In the fifth region, the perfect intentions and aims set for one's SELF are seen. On earth, only imperfect and weak copies of these intentions are realized. He is here really he himself. Only what relates to the eternal progressive course of existence finds its impulse here. He is a member of the Divine World order.

"In the sixth region, he is in the region where he is with the 'Archetypes of what ought to happen according to the right course of the world'.

"In the seventh region, man stands in the presence of the Life kernels."

The above regions interpenetrate each other. Man himself must bring to each the capacity to be conscious of them. The way lies open to modern man who takes the path described in Knowledge of the Higher Worlds and Its Attainment.

"As ice is only a form in which water exists, so are the objects of the senses only a form in which soul and spirit beings exist." (Theosophy 98)

What can an earth-bound mind make out of the above? Fantasy? Hallucination? Weird-oo?

If we got off our high horses and sterile minds and the self-satisfied egotism of "our most highly civilized age in all history" as we are prone to think, and in which all the important questions of man's coming into being, why he is here on "this speck in the limitless universes", and where he is going, or if he is going anywhere at all, are completely unanswered and unanswerable, and consider the foregoing paragraphs seriously, we will gain an understanding of the Reality which lies just behind our sense perceptible world, and learn how the kingdoms of nature and man came into existence.

Look at the pictures of a human nerve cell or neuron as seen through a scanning electron microscope (LIFE Magazine, October 22, 1971) or of other organs, blood system, etc. in other issues of LIFE. The intricacy and elaborate physical structures are unbelievably complicated.

To imagine these wondrous creations to be the result only of physical forces, of molecules coming together willy-nilly, by chance, mutation, 'survival of the fittest' or any such method or non-method is infantile imagining. It pales into insignificance when one compares it to what the Reality is as seen by the true Seer.

The Spiritual Beings who are the Creative Archetypes had worked on the physical forces since ancient Saturn, Sun and Moon to build Wisdom by the Spirits of Wisdom, Movement by the Spirits of Movement and Form by the Spirits of Form. These exist first in spiritual "seed" form as forces, then the etheric (life) forces fill them and finally the materials of earth interpenetrate the forms and make them visible.

Behind the visible heavenly bodies work the

Creative Hierarchical Beings just as behind the human physical body works the ego of man.

THE OUTER PLANETS, THE ZODIAC AND BEYOND

The ego expands farther and farther beyond the sun, into the orbit of Mars, Jupiter, Saturn and beyond into the Zodiac, elaborating the experiences of the last earth life. It is spread out in the Cosmos and is united with the Creative Beings in it. Only the most highly developed human egos take conscious part in the sphere beyond the sun. Less developed human egos take active part only in the inner planets: moon, Mercury and Venus. In the higher spheres, they cannot collaborate consciously and the work of building their future bodies must be done for them by Hierarchical Beings in accordance with what the human egos bring with them from their past life or lives.

RETURN TO EARTH

There comes a point when the expanding ego can make no further progress in the spiritual realms. The ego then has a yearning to return to earth in order to carry out further deeds and to develop stronger faculties and talents which only earth life within a body physically separated from all others, makes possible. It will then be necessary to build a spiritual prototype of a physical body which will include all the fruits of the past incarnations. This is done with the help of the Hierarchies.

From the limits of the world of the fixed stars and beyond, the being of man contracts, becoming smaller and smaller. "He travels through the starry heavens now with more sympathy and in other spheres with less sympathy. As we relate ourselves with each of the spheres we pass through, so will our coming earth life be determined." (Man's Life On Earth and In The Spiritual Worlds 75)

All the hierarchies assist man to metamorphose the experiences of his last earthly life into spiritual forces to form the right physical body for the next earth life in order to carry out the purposes engendered in the soul on the upward path after the last death.

With the help of the Beings in each part of the Zodiac, man acquires their specific forces to build a particular organ of his physical body.

The starry skies are NOT a homogeneous mass of stars, flaming and blazing at higher or lower temperatures, divided into magnitudes, clusters, or spectral class (colors), as astronomers are compelled to do, by the vastness of their numbers. The uniqueness of each planet of our solar system should give our astronomers a clue of the vast differences that could be expected among the stars.

However, what lies behind the physical planet as the Hierarchical Beings and their activities are beyond the scope (both telescope and microscope) of what materialistic science can discover.

The very idea that there could be creative spiritual beings is abhorrent to an intellect that is hypnotized with the idea that atoms, protons, neutrons, etc. can explain all existence.

The modern Initiate who has developed the supersensible organs of perception latent within the most complicated creation on the face of this earth, the living HUMAN BODY, and which NO machine can ever approach in complexity, sees all these Beings! He can confirm and surpass what ancient man with his normal clairvoyant sight saw. Ancient man had one-half of each of the supersensible organs given to him for his use during those ages before his intellect had been developed.

He saw dimly the beings and their forces working from the starry worlds. Man had to lose the use of those clairvoyant organs in order that he might develop his conscious intellect.

Now, in our age, however, through his developed intellect, through his developed thinking, he must develop the other half of each of his supersensible organs. Then he will see clearly as the modern Initiate sees, and not dimly as he did in ages past.

The Initiate sees, for instance, that while the ego is in the Constellation of Leo with the sun Beings sending their forces up to it, the Leo Hierarchies and the sun beings give man the spiritual forces to build his future heart. (Man's Life on Earth and In the Spiritual Worlds 37)

"When you are living in and with those Spiritual Beings in the constellation of Aries, you will work with the Hierarchies of Aries in forming your future head". (Man's Life On Earth and In the Spiritual Worlds 107). How we relate to the Hierarchies of Aries differs in each individual as a result of what we bring to that sphere from our previous earth-life.

Therefore, the spiritual pattern of the head organization of each ego is unique and no two human heads are exactly alike.

How much we derive from any experience on earth also depends on what we bring to it, whether as visitors in a museum, or standing in front of ancient hieroglyphs, a blueprint for a chemical plant or home or atomic power plant, or in the midst of persons speaking in foreign tongues. What we bring with us has an immense influence as to how much we understand and benefit from our experiences in earth-life.

What we derive from the Hierarchies, fortunately, does not depend entirely on our own conscious development, but on what the higher beings determine we need in order to carry our evolution a step further.

The Spiritual Hierarchies in each part of the Zodiac and each planet have highly specialized forces.

The Hierarchical Beings of each part of the cosmos are of the most diversified powers, i.e.:

"While working with the Hierarchies in

Taurus, you elaborate the region of your larynx in its connection with the lungs. While going through the constellation of Leo, the sun forces elaborate the human heart." (Man's Life On Earth 108,37).

The Hierarchies of each of the twelve signs of the Zodiac are the Architects of, and supply the spiritual plan and forces for each and every organ or part of man's body. Man feels spread out in, and lives among these lofty Beings.

In the ego's journey through the zodiac and planetary spheres, the individualized, unique spiritual pattern for the physical body conforming to the ego's needs is assembled and built out of the forces of the entire cosmos.

Each of us is a UNIQUE copy of the relationships of the way how we relate to the Beings of the heavenly bodies when we journey through those spheres during the time between death and a new birth; and also how those Spiritual Hierarchies relate to Hierarchies of other planets, whether supplementing their forces or diminishing them at the time of the ego's passage through them.

This is the true Astro-Logos, the creative Word of the Cosmos of Stars, the Music of the Spheres spoken of by the Ancient Seers who perceived the Spiritual Beings and their activities.

This became the materialistically conceived astrology which gives the impression that the physical matter of the stars influence men's lives, rather than the Spiritual Hierarchies active in those heavenly bodies. In a materialistic age, and in the previous ages when the creative worlds darkened for men's views, astrology has, in general, descended to the level of simply measuring angles and degrees of physical planets and their relationships to each other and to the zodiac, but neglects the Spiritual Beings who are the real influence working from those spheres.

There are some notable exceptions fortunately, among astrologers.

There are infinite variations in human heads, human hearts, lungs, livers and other organs. The spiritual pattern for each organ in the forthcoming physical body is different from every other ego's. A heart is not just the same as every other heart in details, as heart transplant specialists are discovering to their bafflement. Every heart is attuned to and reacts to the forces of the cosmos and more specifically to the sun and Leo Hierarchies in a particular way in accordance with the individual ego's development.

The heart is not a mechanical pump that can be interchanged and replaced with impunity.

The Architectural Plan for the liver, lung, kidney, brain cell, blood cell, DNA molecule, chromosome, etc. is for that ego's special karmic requirement for the next earth-life. A unique spiritual pattern of our physical body underlies every organ and every detail of our body.

In expanding outward after death, the soul and spirit of man in the first period has Imaginative Cognition, then after expanding out to the higher spheres has Inspirative Cognition, and while united with the Spiritual Beings in the sun region, the outer planets and also in the zodiac, has Intuitive Cognition.

After the midpoint between death and rebirth, called the Cosmic Midnight, the ego reverses in cognitive powers, from Intuition to Inspiration and then to Imagination while becoming smaller and smaller and dimmer and dimmer in consciousness. When the ego has contracted to the size of the Saturn sphere, it is in the region where the Archetypes of soul work creatively; in Jupiter, the Archetypes of life and in the Mars sphere, the Archetypes of the physical.

The ego gathers forces also which make him independent in a right way of the forces of the Hierarchies.

After the sun sphere, the ego contracts to the size of the orbits of the inner planets, and the soul world is entered and astral forces are gathered in accordance with the needs of the ego as defined by the spirit orb.

From about the mid-point between death and rebirth, the ego has been working on the physical line of descent, until eventually it reaches the two parents.

In the sphere of the planet commonly called Mercury, the Archangels who are the folk souls of the races and peoples on earth, select the race into which the ego will incarnate, in accordance with the ego's present needs. "In the Mercury sphere, he lives with more or less sympathy for a particular family. A human soul may be descending when Mercury is on the other side of the earth and the soul may be able to disregard those Beings' influence and then have no great connection with his family. Of it may choose to come to earth when the 'ray' that goes from Mercury to that family is best and he can have very close ties with the members of his family". (Man's Life On Earth 77)

In the next sphere, the Angelic Guide of the ego assist in directing the soul to his parents. "The ego draws the person to the paternal part of the parents; the astral body draws him especially to the mother".

It is the "Jahve Being who has his seat on the Moon, who leads man down to earth". (Man's Life On Earth 86) The moon has a condensing, concentrating tendency which draws the ego, spread out among the planets and united with the Hierarchies there, and condenses the spirit orb ever more and more toward earth.

Since the ego started expanding outward from the earth at the last death, and is headed back to earth as earth is his home as a physical being, the earth IS, and must be considered to be, the center of man's cosmos as in the Ptolemaic concept of the universe. This does not negate the fact that the sun is the physical center of the SOLAR system, as there, it is the SOLAR BEING Who has His center of activity.

THE SPHERES THE EGO CONTRACTS FROM TOWARD EARTH-LIFE AND EXPANDS INTO AFTER DEATH

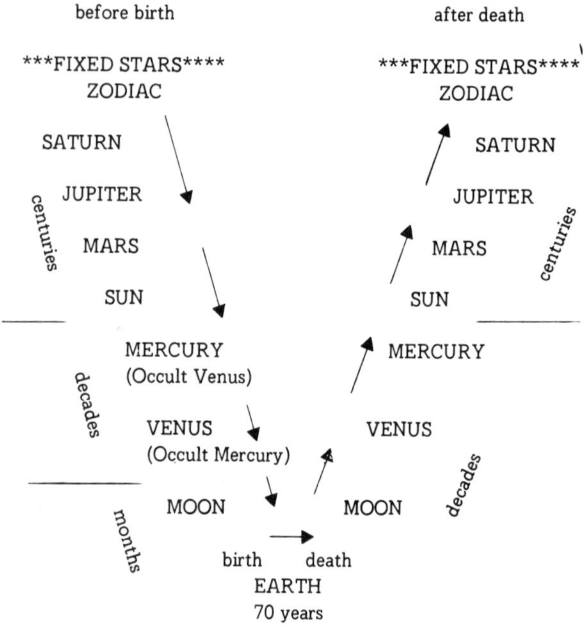

Earth-life is only a small part of the Ego's existence.

Figure 1

"Egos approach the earth sphere mainly from the months of December through March and remain there for shorter or longer periods. Then they remain waiting in the planetary earthly sphere until the opportunity is afforded them." (Reincarnation, Wachsmuth.G. 156)

"It is an important question for the descending human soul to decide whether it shall become man or woman. Its whole destiny on earth depends on the decision . . . . Due preparation must be made. If the soul desires to be a woman, it will approach the earth at the time of the Full Moon. On the other hand, when we, looking from the earth see New Moon, then the soul that is descending can become a man". (Man's Life On Earth 76)

CONCEPTION TO BIRTH

The spirit-germ unites with the physical seed at about the time of conception and enters the stream of heredity. "The spiritual seed of the physical body which we were preparing falls ever farther from us and disappears. We are obliged to witness this: the spiritual seed has fallen from us; it has gone down into a physical mother and father, entering into the forces of generation, into the stream of generation upon the physical Earth. So it is in all reality. The physical body we were preparing shrinks and contracts and falls into the streams of generation,---into a physical father and mother upon Earth,---while we ourselves as soul and spiritual being are left behind, feeling that we belong to what has fallen from us, yet cannot unite with it directly. In this condition---it is our only means of reuniting with it---we now begin to draw to ourselves the forces of the Ether that are there throughout the Cosmos; we begin to form our ether-body. We do this when the spirit-seed of our physical body has already fallen from us and is down there on Earth, preparing the physical body in the mother's womb, while we are gathering the forces with which we form our ether-body. With this etheric body we then unite ourselves, when the human seed has already been for a time in the mother's womb." (Man's Life On Earth And In The Spiritual Worlds, 110/111).

"What until now was but the 'memory' of our own Karmic entity, we now take in as real effective forces, right into our ether-body. Therefore we afterwards appear on Earth in such a way that we of ourselves bring about the unfoldment of our destiny, our Karma."

The ego and astral body have a brief prevision of the coming earth-life just before entering the physical organism within the mother's womb. "Only in the third week after conception do the higher members, the ego-astral-etheric organization, submerge into the process of the physical evolution and enter into the previously accomplished union of spirit-germ and physical germ." (Guenther Wachsmuth, Ph.D. in REINCARNATION, 159, quoting Rudolf Steiner). In the following pages of his book, Wachsmuth shows in a diagram that the spirit-germ unites with the physical germ very near the time of conception and in the third week the ego being, astral body and etheric body descend into the developing entity in the mother's womb, the etheric body becoming more active in the seventh week, the astral body of the child exerting more intense action in the seventh month and the ego of the child exerting more intense action just before birth. (Reincarnation, Wachsmuth, G. p. 161).

In the lectures of the Gospel of St. Luke, Rudolf Steiner says, "When the human embryo develops in the body of the mother, the Ego unites with the other members of the human organism in the third week, but does not come into operation until the last months before birth and then only gradually". (Luke 104)

An angelic being guides the ego to the right parents because the consciousness of the ego had grown dimmer and dimmer as it became smaller and smaller on its return to earth. The spirit plan of man is a tiny spirit-orb containing the forces of the heavens in a unique way, and being a replica of the zodiac and solar cosmos in the special relationship as to how their forces work on his entire make-up.

Out of the potentially large number of combinations and variations possible within father and mother, the particular chromosomes, DNA and genes matching the forces in the spirit-orb are

THE UNION OF MAN'S HIGHER MEMBERS WITH THE PHYSICAL ORGANISM IN THE MOTHER'S BODY

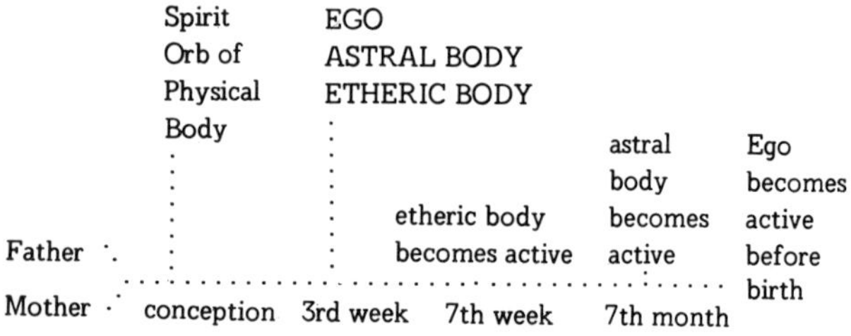

Figure 2

attracted and drawn together by the forces in the spirit-orb that has condensed from the limits of the cosmos to form the physical germ.

Father and mother, according to our scientific view, are the sole creators of the child. To the eyes of the Seer, parents are the last step in a long preparation that stretches from beyond the limits of the zodiac and contracts nearer and nearer towards earth. They are the means for bringing the spiritual pattern as closely as possible into a physical body.

"Man carries within him infinitely more secrets and mysteries than the worlds we can perceive with our external senses and explain with the sense-bound intellect". (Man's Life On Earth and In The Spiritual Worlds 60).

"From the human germ cell, even if it were fertilized, nothing could result that might live on earth if the compressed spirit-globe could not unite with it. From it only the foundation for the senses and the nervous system could result, but nothing which is capable of living, such as the body of man which must build itself up around the senses and the nervous system. Earth can give the forces for the senses, the nervous system. What grows organically around them, must be brought in from the cosmos". (Occult Research Into Life Between Death And A New Birth 16)

Can our embryologists and biologists accept this fact? Most likely not. So they will attempt to inseminate ova in test tubes. Most will die quickly or develop senses and nerves, but if some Ahrimanic being should decide to unite and inhabit such an organism, it would not be a normally developed human being.

Will our scientists then be creating an ogre or a demonic monster? And will they then come out with the excuse: "Oh, I didn't know such a horror would develop. Nobody told me. I was only doing what somebody asked me to do, and I agreed it would be an exciting, necessary and interesting experiment."

Just the way it happened with the atomic bomb and the H-bomb?

The ideas of heredity about our coming into existence at birth pales into insignificance, inadequacy and childishness, compared to the Reality seen by the Seer or exact clairvoyant of our age. Man's physical body is the work of the Mighty Hierarchies Who serve God! Man's physical body is truly a Temple of God because the ego is a drop of the Cosmic Ego and the human body is a copy of the starry worlds which is the 'body' of the Hierarchies behind all of whom is God.

ON ABORTION

The materialists take the view that up to eighteen weeks, or five months or even six months after conception it is only the mother's concern if the life of the unborn is terminated. That is based on the view that only the forces of the parents have brought the child into being. Actually the ego of the unborn and the spiritual hierarchies of the cosmos have worked hundreds of years at preparing the spiritual blueprint for the developing body. And these have united with the embryo as early as soon after conception in the case of the spirit-orb or spiritual pattern of the physical body; in the third week in the case of the ego-astral-etheric; the child's etheric body exerting more intense action in the seventh week; the astral body in the seventh month and the ego just before birth.

OTHER COMMON ACTS OF MAN

How does man treat his physical body? He poisons it by overeating foods treated with toxic sprays of inadequately tested chemicals; he devitalizes what is left by over-processing, even removing the most essential elements and then adding others less nutritious. He then adds other chemicals "to retard spoilage" which prevent even the hardy mold from forming for weeks and months. One wonders whether the insides of man would be 'retarded from spoilage' also from ingesting such food.

He further poisons his body with alcohol, and uses his lungs as an unrenewable filter for smoke containing the products of combustion of cigars and cigarettes. Then he dulls and destroys it with drugs, LSD, marijuana, and opiates. Instead of a thinking being, he becomes dulled into insensibility. He destroys what took the Hierarchical Beings hundreds of years to bring into this incarnation, and millions of years to bring to this stage of evolution.

In hypnosis, he callously inserts another ego into a human organism specially prepared for hundreds of years just for that original ego-inhabitant. How the hypnotist can upset the ego-astral-etheric-physical organization and the hypnotist's karmic responsibility for having done so, he will discover after death. He callously murders by legalized abortions carried to extreme degrees. He degrades by living for pleasure in sordid conditions.

He murders and thinks non-detection or clever arguments to escape punishment forever releases him from the consequences of his crime.

"The descent to birth is a transformation of the wisdom of the universe into the life-forces of the earthly structure. Only through transforming the wisdom, which flows to us abundantly, into life-forces do we approach a new incarnation in the right way. And when we come again to the earth, we must have transformed so much wisdom into life-forces that we are able to permeate with enough organizing spiritual life-forces the hereditary substance that we receive from father and mother."

The hereditary streams may not provide the ideal organism for the spirit. Certain adjustments are made by means of childhood diseases, and the spiritual worlds work forcefully into the child, especially in the first three years before the child has a consciousness of self. This is the point to which our memory can go back to in later life. Before this period, the child refers to himself in the third person: "Tim did that", rather than "I" did. In those first years of life, the miracle of gaining the vertical position (against the forces of gravity), acquiring the power of speech, and organizing the brain as an instrument for thinking takes place in close cooperation with the Spiritual Beings who are close to the child. The Seer can observe this as the aura which surrounds the child. After gaining self-awareness and being able to say, "I", the higher beings withdraw and the aura which extended around the child draws within him. "The wisdom that works on the child in its first three years of earth-life is never again as high after he becomes conscious of himself as an earth-being."

The above short sketch of the soul and ego's progress from death through the soul and spirit worlds and back again to earth-life must sound fantastic to a materialist, or even to the most religious person.

Consider the fact that from a quarter million miles away, our astronauts while on the moon or circling it, see earth as a blue planet floating in space. The billions of people on earth are not seen by them. Our existence would not be suspected by a stranger approaching earth from the distance of the moon, which is a mere stone's throw away in terms of cosmic distances.

How mistaken we can be about planets and stars immeasurably distant especially when we do not consider the invisible worlds a reality, and do not even know about the invisible members of man's being.

"MAN! KNOW THOU THYSELF" sounds as a cosmic directive more urgent in this age than any other age before us. Our potential for evil exceeds that of any other age. We have become freed from the dictates of religion and of religious leaders by our predominant idea that each person knows just as much as anyone else in regard to man's origin and destination in life; and the dominant idea in our time is that NO one knows anything about it. How mistaken and arrogant this idea is, few suspect.

If we study the above paragraphs and the books referred to with an open mind, we will gain an understanding of the secrets of life hidden behind veils impenetrable to materialistic science.

We will get a glimpse of man as the THOUGHT AND DEEDS OF GODS carried out over unimaginably long epochs, OF MAN AS THE PURPOSE OF ALL THAT SURROUNDS US AS THE SOLAR SYSTEM, OF MAN AS THE TENTH HIERARCHY OF SPIRITUAL BEINGS, NEXT TO ANGELS, ARCHANGELS, ARCHAI; SPIRITS OF FORM (ELOHIM); SPIRITS OF MOTION (DYNAMIES); SPIRITS OF WISDOM (KYRIOTETES); SPIRITS OF WILL (THRONES); SPIRITS OF HARMONY (CHERU-

BIM); and SPIRITS OF LOVE (SERAPHIM) who are the servants of the TRINITY (GOD THE FATHER, GOD THE SON and THE HOLY SPIRIT).

CHAPTER IV

Creation

What does the Akasha Chronicle reveal about the creation of our solar system? Rudolf Steiner in his books OCCULT SCIENCE AN OUTLINE and COSMIC MEMORY as well as other books, gives the results of his researches in the eternal spiritual records. He purposely calls the first book "an outline" in spite of its 300 plus pages. What follows in the next few pages can be at best, the merest indication of the vast vistas he reveals in those books, of the coming into being of man and the kingdoms of nature as well as of the various planets of our solar system.

The creation of our solar system started as a spiritual activity. It predates anything known to science.

In the beginning, from out of the great universe, or more strictly speaking, the zodiac, nine hierarchies of Spiritual Beings were summoned from the realm of the Father God, by God the Son to form a new hierarchy, MAN. The highest two hierarchies, Seraphim and Cherubim, or Spirits of

Love and Harmony, 'stand in the Presence of God' and interpret God the Son's intentions. The next highest hierarchy, the Thrones or Spirits of Will had reached a point of development where they were able to sacrifice of their substance, Will. This is of the nature of the warmth of soul such as is fired in the enthusiasm of a person today.

ANCIENT SATURN

This warmth is spiritual in nature and is invisible and not connected with substance, because there was no substance in our solar system at that time.

The warmth was worked on by the other hierarchies to form a spiritual beginning of the first rudiment of the physical bodily forces of man. Occult Science, i.e., that science hidden from the ordinary gaze of man, calls the development of the warmth sphere Ancient Saturn, or simply Saturn. These terms and their connection with our present planets bearing the same names will be made clear later.

"Saturn consisted in the beginning of nothing but the very first rudiments of humanity . . . Man existed at that time only in the first rudiments of his physical body. Ancient Saturn was simply composed of individual physical human bodies". (The Influence of Spiritual Beings Upon Man 38)

The development passes through seven stages, each succeeding stage worked on by the next lower hierarchy, the Spirits of Will, Wisdom, Motion,

Form, Personality (Archai), Archangels and Angels. While working on this member of man's being, each hierarchy in the course of Saturn evolution advanced a step. On Saturn, the Archai go through their humanity stage or ego stage, with warmth as their densest form.

Not all the beings in each hierarchy and 'man' reach full development. There are those which fall behind. At the end of the Saturn evolution, the laggard human entities form the rudiments of the present animal kingdom.

Human consciousness was of a deep trance condition. His entity, however, was used by the hierarchies to express themselves in their creative activity.

At the end of Saturn, the Spirits of Will (Thrones) gave the predisposition for the highest member of man, spirit man.

A cosmic 'sleep' state follows in which all is united. For a fuller account, see Cosmic Memory, Chapter 14, or Occult Science.

THE SUN EVOLUTIONARY STAGE

A second planetary evolution called Ancient Sun or simply Sun, begins. A short repetition of the Saturn development takes place to prepare the physical organism of man to incorporate another member.

The next highest hierarchy, the Spirits of WISDOM (Kyriotetes or Dominions) sacrifice of their substance to give man his etheric body! This is the "gift of mobile and power-filled wisdom, the gift of LIFE that is bestowed by the Spirits of Wisdom." The whole of the Sun Sphere was filled with LIFE! Life was the gift of the Spirits of Wisdom. Warmth ether condenses into 'air' as its lower part and into light ether as its more refined part. "The Spirits of Motion then implant a powerful activity in it. The Spirits of Form give the two-membered entity of man a definite form. He becomes a human-plant. His consciousness is that of dreamless sleep. One can compare it with the dull consciousness of plants of our day." (Cosmic Memory 184)

The Archangels or Fire Spirits go through their humanity stage on Sun, with 'air' or the gaseous element as their densest form and using the two-membered human entity for its field of activity. The physical body advances in its development and "the Spirits of Wisdom bestow wisdom on the limbs to render their motion a wisdom-directed one." (Cosmic Memory 188)

On Sun, the physical body goes through its second stage of development, becoming more perfect thereby, while the etheric body goes through its first stage.

At the beginning, Sun was a sphere the size of the orbit of our Jupiter around our sun. In it was LIFE. In the later stage there were two planets, the sun acting external to the remaining body.

Towards the end of Sun development, the Cherubim pour out their wisdom on the etheric body to give man the disposition for future development of his life-spirit. There follows another cosmic rest period in which everything unites into one body. Each hierarchy rises another stage, while "others fail to develop fully and some become the Spirits of Darkness which implant independence and selfishness into man." For further details, see Cosmic Memory Chapter 15 or Occult Science.

THE MOON EVOLUTIONARY EPOCH

After a cosmic sleep period, a third planetary development takes place called Ancient Moon, or Moon epoch. From a unified body, a short repetition of the Saturn and Sun development takes place to prepare the plant-man to be able to take into itself a third member. A further condensation to water, or the liquid state, takes place and a higher etheric force, chemical or sound ether develops.

The next hierarchy towards man, the Spirits of Motion or Dynamies reach a point in their development where they are able to sacrifice of their substance to form man's astral body. "The Spirits of Form shape the astral body so that its whole organization can develop inner processes. These have the character of man's instincts." (Cosmic Memory 199)

"Man consisted of warmth ether, air and water as his densest form and in the later stages, the skeleton of man was cartilaginous." (Cosmic

Memory 205)

"The angelic hosts inhabit the human physical body and go through their humanity stage within the liquid state densifying as far as the cartilaginous."

The ancient Moon sphere was the size of the orbit of Mars around our sun. "Man is in a state of consciousness similar to that of dream-filled sleep. It is an image consciousness, symbolical expressions for the events in the dreamer's environment." (Cosmic Memory 196) "At a certain point, the Moon splits off from the unified Sun and revolves around it so that man can develop at the slower rate on the remaining Moon body."

The developing astral body is inoculated with independence by the Spirits of Personality. In man's etheric body a kind of joy and pain develop and in his astral body, rage, hate, instincts, passions and so forth. The entity of man is animal-man."

"The Seraphim, in the later Moon epoch, gave man the possibility to develop spirit-self," the third member of higher man.

"In the Moon Epoch, the Spirits of Wisdom incorporate wisdom into the whole of creation. Wisdom is built into all the forces and currents which underlie every organ and cell in man, and into the animal and plant kingdoms as well. Ancient Moon is therefore known as the Cosmos of Wisdom." Therefore, we on earth find wisdom in all of creation, from the smallest to the largest.

At the end of the Moon epoch, everything is reunited in one mass and sun and the remaining Moon body become one body. A cosmic sleep period follows.

EARTH EPOCH

A fourth cosmic condition called Earth begins with everything which had previously developed existing in a unified mass as in a spiritual 'seed'. The condition is as if the previous Moon condensation to as far as a gelatinous state had dissolved, rarefied and become spiritual warmth again. One might imagine mushy snow floating on water, dissolving and the water vaporizing and disappearing from sight.

This spiritual warmth with the Spiritual Hierarchies weaving and streaming in creative activity within it is the beginning of our present solar system. "It is of a size extending far beyond Uranus and the outermost planets belonging to our solar system." (The Influence of Spiritual Beings Upon Man, 39).

THE EVOLUTION OF THE SOLAR SYSTEM

From a spiritual non-material state (pre-gaseous nebula) Old SATURN, condensing to the gaseous Old SUN, to liquid Old MOON, to our present solid EARTH state.

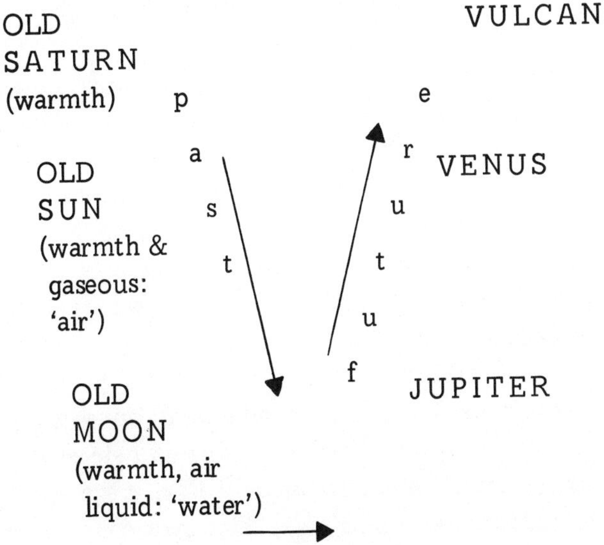

Present Solar System

EARTH
(warmth, air, water,
solid state: 'earth')

Future development takes on more rarefied conditions: JUPITER, VENUS and VULCAN states as man evolves higher.

Figure 3

THE OUTER PLANETS: SATURN, JUPITER, MARS

A short, higher cycle of Old Saturn evolution takes place. It was now necessary that the form and figure of man be changed so that objective consciousness, instead of the previously developed pictorial consciousness, become possible. At the end of the preceding Old Moon evolution, the physical body was adapted to only a picture consciousness.

The higher Beings work on the physical bodily forces of man alone. They transform or metamorphose the forces of the physical body of man for the fourth time. Wisdom is inherent since the end of Old Moon. The physical body, condensing to as far as the astral in this period, becomes the most perfect of the members of man's being.

At the end of this cycle, certain beings who had remained behind in the old Saturn stage, leave the unified Earth body and take their substances with them, to form the beginning of our planet Saturn. There they can best carry on their activity in a warmth ether state.

"In this first period, there is only the human realm existing as physical forces. The sensory organs, eyes, ears, etc. have their origin in this period." (See Occult Science and Cosmic Memory for details). At the end, all the Beings pass into a higher form of existence.

In a second small cycle, a short repetition of Old Sun follows on a higher level, during which the etheric body is worked on by the Hierarchies, and undergoes its third metamorphosis, preparing for the entry of the fourth member of man's being. "Organs of growth, nourishment and reproduction have their beginning. The physical human body is a system of glands." (Cosmic Memory 231).

In this period other spiritual beings who cannot work in the further densifying fire-air state, leave the remaining Earth and form the beginning of our present planet Jupiter.

The fiery-gaseous condition of the old Sun recapitulation is the earliest condition to which physical astronomy can theorize as to the origin of the solar system, but they cannot reach into the activities of the Spiritual Hierarchies who were fashioning the foundations of man's physical body and etheric body at that time.

In the third small cycle, a short higher repetition of Old Moon follows. The astral body is added by the Spirits of Motion who guide its second metamorphosis. The first rudiments of a nervous system is integrated into the evolving organism by the Hierarchies. The entities of man divide into two groups.

"Higher man and lower man followed separate paths at first. Lower man, at the repetition of the Moon cycle, is a being midway between man of today and an animal. The plants stand midway between the animal and plant nature of today and the minerals only half bear their lifeless character of today, while for the rest they are still half plants." (Cosmic Memory 211).

Other spiritual beings who cannot work in the further densifying warmth-air-water state, separate from the remaining Earth and take their forces and substances along to form the beginning of our present planet Mars.

Later in this cycle, the Fire Spirits and higher beings separate the sun from the remaining Earth body. These higher Beings cannot work at the slow tempo connected with Earth of this period. The astral body can then become imprinted on the human entity.

"During the last half of this third cycle, the minerals harden, the plants gradually lose the animal character of their sensibility and out of the uniform species of animal man, there develop two classes. One of these remains on the level of animality while the other is subjected to a division of the astral body into two parts. The lower part of the astral body continues to be the carrier of the emotions, while the higher part attains a certain independence so that it can exercise a kind of mastery over the lower members: the physical body, etheric body and the lower astral body."

The spiritual hierarchies work actively in these members of man and it is their activity which produce wisdom, movement, form, personality, etc. in the physical body, etheric body and astral body throughout the cosmic epochs. These members gain greater and greater perfection and development so that they can be used by man.

After each of the three small cycles repeating on a higher level, old Saturn, Sun and Moon cosmic epochs, a reuniting of all the separate elements takes place and a more spiritual condition follows, a condition not accessible to ordinary seership.

EARTH CYCLE

A fourth cycle having seven epochs of development begins. It is necessary to repeat the previous conditions in a shorter period before evolution can proceed further.

The first is the Polarian epoch. In this cycle, "Earth consisted of warmth ether. In this substance, man existed as a fire-mist. It is as dense as a cloud suspended in the air. Fire-cloud man is internally alive and organized; man is in soul-slumber and only dimly conscious. He is directed by higher beings." (Cosmic Memory 133).

SEPARATION OF THE EARTH FROM THE SUN

In the second epoch, called the Hypoborean epoch, the Spirits of Form, together with the higher hierarchies, eject the densifying earth-moon forces and substances, and retain our present sun as their field of activity. These lofty beings would have been impeded in their activity by the earth-moon substances. Also, the human entities could not develop at the rapid rate of the sun forces and would have been quickly consumed by them. The ejected Earth in that period after the separation from the sun consisted of warmth (fire), air and water.

THE LEMURIAN EPOCH

Man continued his development on the remaining earth-moon. "The sun forces now worked on him from outside, while previously they worked on his members from within while sun and Earth were still united."

As the earth-moon body became denser, "the human substance now to be found there became too hard and ligneous to enable them to incarnate." (Luke 82) Fewer and fewer human souls were powerful enough to utilize the hardening earthly bodies. Finally only "one main pair retained sufficient strength to master the stubborn substance and to incarnate on the Earth . . . through the period when the Moon was separating from the Earth." (Luke 83)

Spiritual Science independently confirms the Adam and Eve account of the bible.

The other human souls foresook the earth-moon and formed spiritual colonies on the planets: Saturn, Jupiter, Mars, Mercury and Venus.

It became necessary to remove the densifying forces in the earth-moon which was still in a warmth-vapor-fluidic state. Fiery volcanic forces were active on the continent of Lemuria where this development was taking place. "Ancient Lemuria was a continent between the present continents of Asia, Africa, and Australia, and was inhabited by a very different humanity."

SEPARATION OF THE MOON

Towards the middle of Lemuria, the seventh leading Spirit of Form, at a tremendous sacrifice of not continuing his normal development on the sun with those of His Hierarchy, "chose another task for the benefit of humanity. This being took out the densest of the earth-moon forces, together with the matter connected with those forces and separated our moon out, leaving our present earth."

"This event caused a great revolution. The objects which surround man lost a great part of their heat. These objects thereby entered into a coarser and denser substantiality. Man must live in this cooler environment." (Cosmic Memory 136).

The separated moon forces would have hardened the human entities and would have prevented his development in the right way. This Spirit of Form thus works on man from outside since then, just as the sun beings do. This Being is the Elohim known as Jehovah in the Old Testament. Jehovah is the Regent of the Moon who guides mankind's development from that sphere.

"After the separation of the moon, substantiality increases. Rudiments of hands and feet form on the human entities on earth." The higher Hierarchies work creatively in the entity of man, transforming his members.

"During the Lemurian epoch, man assumed a form which made it possible for him to develop his fourth principle, the Ego."

The Ego was the gift to humanity of the six leading Spirits of Form who had made the sun their sphere of activity. They sacrificed their substance, the Cosmic Ego, and gave man a "drop" of this ego. The totality of these six Sun Beings is called in our age, CHRIST.

"At that time, the first seed began to form for the development of the ego in the other three principles. Before the Lemurian epoch, the Earth was also inhabited, but by human beings who as yet bore no ego within them. They consisted of the principles which had been brought over from their former development during the planetary evolution of old Saturn, old Sun, and old Moon. These human beings consisted of physical body, etheric body and astral body." (Luke 81)

"Before the separation of the moon with its densifying forces, no ego could live in the human bodies on earth."

The strongest pair of souls remaining on earth when the moon separated were the ones into whom the divine Beings of the sun first gave the ego.

"After the moon separated there is a transformation of the human form. Those organs which previously had served for nutrition and reproduction are transformed into organs of speech and thought. Man has become upright." (Cosmic

EARTH EVOLUTION
FORMATION OF THE PLANETS

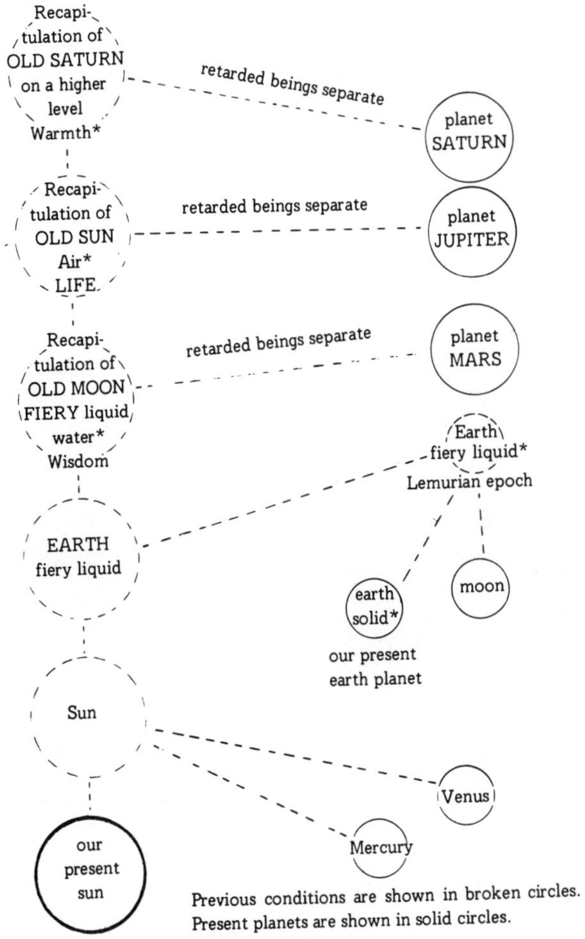

*densest form

Figure 4

Memory 136/7)

EARTH

After the separation of the moon, earth as we know it today began. "After the moon's withdrawal, the entire plan of evolution appears in reverse." (John 61) Before the separation of the moon, there was on Old Saturn, man; on Old Sun, the animals came into existence in a rudimentary stage; on Old Moon, the plant kingdom was added; finally on Earth, the mineral kingdom precipitated.

After the separation of the Moon, there was a reversal in the appearance of the kingdoms on earth. The mineral kingdom revivified first, then the plant kingdom. The animal souls also hastened to unite with the earthly bodies. They did so too soon as the bodies were too hardened and had not been sufficiently revivified by the sun forces, and the animals became fixed in their forms. The astral qualities or instincts worked in a moulding way on the astral currents, then on rarified air, then on the cloud-like-fluidic substances. Finally when the earthy materials precipitated out, the forms became fixed because the instincts united with the physical bodies. The astral qualities and instincts determined the physical form of each animal species. Into these had been built all the wisdom in the Old Moon period by the Spirits of Wisdom.

Study each animal species and see how well

its inner quality is expressed in its form. The astral qualities were already present on Ancient Moon before there was solid matter. Only on earth is there the solid state of matter and this permeated the astral forms to produce the visible animal species.

These astral qualities are in reality exudations of the human astral qualities developed on Moon. They take on separate physical form, thus freeing human beings for higher development.

The animal kingdom, therefore, are those beings which take out from humanity numerous retarding astral qualities and concentrate them within the animal species. Mankind therefore, owes to the animal kingdom the possibility of developing to higher stages unhindered by all the hampering, retarding astral qualities precipitated out by the animal kingdom.

How does mankind repay the animal kingdom? He slaughters them and eats them, he hunts them for their fur, hide, horn, tusks and blubber, or decimates them for his pleasure and 'sport'.

Are we human or have we become subhuman?

The human souls waited longest before returning to earth. They waited until bodies had developed to a sufficiently high level for their use before descending from the planetary spheres. Then they gradually returned from the other planets as bodies plastic enough for their development could again be formed on earth. These returning souls grouped themselves in oracles known by the names of their planets, as Saturn Mysteries, Jupiter, Mars, Mercury, Venus, etc. All these were found in the ancient world and were the beginning of the races.

The human form up to the end of Lemuria was still very primitive seen from a physical standpoint. Creation is an inconceivably long process. Eventually the fire storms brought on by the degenerating forces spawned by the developing beings destroyed the continent of Lemuria.

The regions occupied by the "European, Asiatic and African peoples were mostly still under water. However, a great continent stretched across the area now covered by the Atlantic Ocean: this was ancient Atlantis." (Luke 80)

Before the end of Lemuria, the most advanced inhabitants were led over the Northern part of Africa to Atlantis.

THE ATLANTEAN EPOCH

"Until far into the Atlantean age everything was embedded in masses of water-mist." (Biblical Secrets of Creation 110) This thick water-mist hid the sun from the eyes of early Atlanteans. "It was only toward the middle of the Atlantean period that this human creature was dense enough to become flesh and bone. At the same time, the earth also became more solid and the human being then descended upon the surface of the earth." (John 1908:103)

This statement gives a hint of the expanse of

time that man developed on Atlantis. For details see Occult Science, Cosmic Memory, Atlantis and Lemuria by Rudolf Steiner.

Atlantean men were led by Initiates who had special relationships to the sun Beings or to the Saturn, Jupiter or Mars Beings and also to those of Venus and Mercury, and of another Leader called Vulcan. (See Occult Sci. 194) In the Oracles, the Initiates trained leaders from among the most advanced human beings.

Man lived more under the guidance of a group soul than as an individual ego. He came also under the influence of lower spirit beings who had not kept up with their respective groups and now worked on the human entities in a tempting way. These beings were under the influence of a leader known as Lucifer.

Since men were less closely united with their bodies in early Atlantis, they had "the faculty to place at their own disposal the forces of growth and reproduction in animal and human nature." (OS198)

"A widespread corruption of mankind ensued. As even some of the Initiates succumbed to the temptation of the lower spirits, the evil grew to greater and greater dimensions. The forces of growth and reproduction when torn from their mother-soil and independently employed stand in a mysterious relationship to forces that work in air and water. Mighty and ominous powers of nature were thus let loose by the deeds of men, leading eventually to the gradual destruction of the whole territory of Atlantis by catastrophies of air and water. As a result of the great storms, the whole face of the earth was changed. Europe, Asia, Africa and America began gradually to assume their present shape . . . Vast numbers of human beings migrated to these countries."

POST ATLANTEAN CIVILIZATION

The leaders of the Oracles (both good and evil groups) led the migrating masses from Atlantis. One of the most advanced groups was led north and eastward by the "sun Initiate who gathered around him men who always included the most direct descendants of 'Adam and Eve', the first ancestral pair who had survived the Earth's winter." They travelled all the way to the Gobi Desert and eventually sent the most advanced portion to India to start the first post-Atlantean civilization. This Indian civilization pre-dates the one known in history by about four to five thousand years, as no written records were made until the time of the third post-Atlantean culture.

Ancient Indians still had a deep insight into the spiritual world and looked to it as something they wished to escape to from life on earth. The seven Holy Rishis were the great teachers of that period. The sun Initiate was the most advanced, while the other six held a position similar to that of our planets to the sun.

Atlantean civilization developed in seven

epochs. The above group reaching India was mainly the fourth civilization of Atlantis, the middle of the seven epochs. Such a middle culture could still make a decided change in the direction of their development in order to introduce a new course in human evolution.

The fifth, sixth and seventh cultures carried Atlantean development to its ultimate point. The seventh group migrated westward and we see the wisdom of the past preserved in Chinese civilization, even when European civilization was still groping in lower stages of development.

Other groups migrated to America, both north and south. The temples we see unearthed in the western world had their origin in the oracles and temples of Ancient Atlantis.

THE PERSIAN EPOCH OF CULTURE

As one culture developed to a high level in an ascending direction, it reaches a certain summit about the middle of its epoch, then there is a leveling off, and in the last third of the epoch, there is a decline. "Traditions preserved in humanity become antiquated and require a new impetus. Rejuvenation is necessary." (Luke 84) Another group of human beings then carry human development up to another plateau.

The people of the Persian culture were more concerned with life as earthmen. They applied themselves to acquiring a knowledge of the sense world, as their perception of the world of spiritual beings faded more and more. The leading Initiate from the Gobi region sent a high initiate who still had far-reaching clairvoyant powers. He taught his people of the Sun Spirit whom he named Ahura Mazdao. He also taught them of the adversary whom he named Ahriman. This leader was Zoroaster or Zarathustra, again not the later one known in history, but the original one: nothing was written down until the third epoch of culture. So the Zarathustra known in history is one who bore that "title" or name in a later period.

Zarathustra taught selected pupils as well as the whole people. We shall learn more of his two leading pupils later.

"Zarathustra taught men of the God <u>without</u>, i.e., in the great Cosmos. He taught men to apprehend the great Cosmos spiritually . . . It was his mission to inculcate into men an attitude concerned with the phenomena of the external world, to present a view of the Universe to a people whose task was to labour in the outer world . . . with energy, vigour and efficiency and with a knowledge and certainty that man rests in the bosom of a divine-spiritual world." (Luke 97-)

THE THIRD POST-ATLANTEAN EPOCH OF CULTURE

Following the Persian Epoch is the Babylonian-Chaldean-Egyptian cultural epoch. Zarathustra gave his astral body to one of his leading pupils of the previous epoch to whom he had taught sun wisdom. This pupil inaugurated the Egyptian civilization. He was called Hermes-Trismegistus, the thrice-great Hermes. The astral body contains the consciousness and knowledge gained in all his many incarnations.

Zarathustra gave his etheric body to another of his leading pupils to whom he had taught the wisdom of earth in the Persian epoch. This pupil reincarnated as Moses who saw the Sun Being approach earth and live in the elements of earth, in the fire and lightning of Mount Sinai, and in the burning bush. Moses became acquainted with the Hermetic wisdom, but "this wisdom had become very decadent and degenerate at the time of Moses due to practices in black magic, which is putting occult powers to use for selfish purposes."

"Zarathustra was reborn in Chaldea six hundred years before our era (at the time of Buddha in India) and worked there as the great teacher 'Nazarathos' or 'Zaratas', who was also the teacher of Pythagoras . . . Since the days of ancient Persia, he had become more and more mature." (Luke 86)

THE FOURTH POST-ATLANTEAN EPOCH

The fourth post-Atlantean epoch of culture was the Greco-Roman which started in the eighth century before Christ and lasted until the fifteenth century. In this period, mankind was approaching darkness even in the soul world. Life on earth was the only world for man: that was the inspiration of the Ahrimanic beings.

Man developed the intellectual soul in this age. Aristotle taught and left a brilliant trail in the ancient sky.

Western civilization spread throughout the known world to as far as Asia in Alexander the Great's conquests. These had a unifying influence on world civilization.

Alexandria became a world cultural center and passed into oblivion.

The rise and fall of the Roman Empire occurs within this period. The legal system was developed.

The Scholastics and the Dominicans were the dominant figures in the eleventh to the thirteenth centuries.

It was during this period that the cosmic event that gave meaning to earth and cosmos took place. This was the CHRIST EVENT.

PLANET EARTH
After Separation of the Moon

LEMURIAN EPOCH

Development on continent of Lemuria lying between Australia, Asia and Africa.

Man is endowed with his fourth member, his Ego.

ATLANTEAN EPOCH

Development on continent of Atlantis lying between Europe, Africa and America.

POST-ATLANTEAN EPOCH

1 Ancient Indian 7227 B.C. to 5067 B.C.
2 Ancient Persian 5067 B.C. to 2907 B.C.
3 Babylonian-Chaldean-Egyptian 2907 B.C.-747 B.C.
4 Greco-Roman 747 B.C. to 1413 A.D.
5 Our present epoch 1413 A.D. to 3573
6 Sixth cultural epoch 3573 to 5733
7 Seventh cultural epoch 5733 to 7893

JUPITER OF EARTH

VENUS OF EARTH

Table 4

CHAPTER V

Jesus and The Christ Being

Biblical scholars have written exhaustively about the discrepancies of the Bible and about the different details in the four gospels of St. Matthew, St. Mark, St. Luke and St. John. Doubts have been raised about the divinity of Jesus and of Christ. Many ministers are preaching about "the man Jesus" and many writers refer to Jesus as "a good man", just another man like the rest of us.

What does the Spiritual scientist find in the Akashic Record?

The following are some pertinent points gleaned from many lectures by Rudolf Steiner on the subject, and the reader will be given direct references in the texts so he can study more fully what is indicated here in a sketchy manner.

THE JESUS CHILD OF THE LUKE GOSPEL

The Gospel of St. Luke traces the geneology of Jesus back to " . . . the son of Nathan, which was the son of David." (Luke 3:31) Nathan was the high priest, one of the two sons of King David. The other son was King Solomon. This Jesus Child is from the priestly line of descent and can be identified more precisely as the Nathan Jesus. The body of the Nathan Jesus is the one which eventually goes through the crucifixion.

The geneology then continues backward to its beginning: " . . . which was the son of Adam, which was the son of God." (Luke 3:38)

The Nathan Jesus is traced back to Adam, and Adam is the son of God. How are we to understand this? Rudolf Steiner's researches in the Akashic Record discovers the following:

In the Nathan Jesus Child incarnated a part of the "mother-soul of humanity" which was kept young and pure just as it was when first given to humanity by the Cosmic Ego in Lemurian times. One part had incarnated in Adam. The Nathan Jesus therefore was the son of God just as Adam was. This pure soul had been kept out of the stream of earthly incarnations. It was therefore untainted by Luciferic temptations. This pristine, innocent soul retained all its youthful forces; its fresh, rejuvenating forces.

St. Luke continues: "And suddenly there was with the angel a multitude of the heavenly host praising God and saying". (St. Luke 2:13.) Who does the Akashic Record reveal the "heavenly host" to be? "The 'heavenly host' was the picture of the glorified Buddha, seen by the shepherds in vision;---We look with the shepherds into the manger where Jesus of Nazareth was born, and see the radiance above the little Child. This Child--was to be shone upon from the very day of birth by the power radiating from Buddha in the spiritual world." (Luke 54)

When the Child Jesus was presented at the temple, "there was a man in Jerusalem, whose name was Simeon . . . . . ; and when the parents brought in the Child Jesus . . . . . he took him up in his arms, and said, "Lord, now lettest thou thy servant depart in peace, according to thy word . . . For mine eyes have seen thy salvation . . . . "St. Luke 2:25-30) Who was this Simeon? "In India long ago---when the Buddha-to-be was born as Bodhisattva--the whole momentous significance of the events was revealed to a sage--Asita was his name--to go to the palace to look for the little Bodhisattva-child . . . . Then Asita began to weep and when asked why, said, 'I am weeping because I am so old that I shall not live to see the day when this Savior, the Bodhisattva, will walk the earth as Buddha.' But the same Asita was born again as the personality who, in the Gospel of St. Luke, is referred to as Simeon in the scene of the presentation in the temple. Having 'the Holy Spirit upon him' he was able to perceive, at the presentation in the temple, the radiance of the glorified Bodhisattva above the head of the Jesus-child of the House of David. Then he could say to himself: Now you need no longer grieve, for what you did not live to see at that earlier time, you now behold: the glory of the Savior shining above this babe. Lord, now let thy servant die in peace!'" (Luke 55)

This great Teacher of mankind had taught humanity COMPASSION and love in his many incarnations as a Bodhisattva until finally in the incarnation in the sixth century before our era, he could fully descend into a human body through which his powers could take effect." (Luke 110)

The Bodhisattva thus attained Buddhahood, a state in which he henceforth can further human evolution from the spiritual world.

"Buddha imparted to mankind what the human soul can find as its own law and obey (it) in order to purify itself and thus reach the highest level of morality attainable on Earth. Buddha was the first to reveal it." (Luke 112)

His deed made it possible for human beings to develop compassion and love out of their own being. Christ was THE BEING of Whom Buddha taught: Christ was the Being of Love itself.

Buddha in his glorified astral body "overshadowed' and hovered over the Nathan Jesus Child nearly 2000 years ago, uniting his mission with Jesus. His earthly followers, even in our day, consider and keep only what he had taught 2600 years ago.

Mankind needs to know these secrets of existence and must not remain bound in the tradition which was suitable for his development in his previous incarnations. These traditions would become harmful if retained rigidly in the old form, and acted to keep men separate.

THE JESUS CHILD OF THE MATTHEW GOSPEL

The Gospel according to St. Matthew gives the geneology of Jesus beginning with Abraham to "David the King; and David the king begat Solomon". (St. Matthew 1:6)

If one compares the geneologies given by Matthew and Luke, from Joseph back to David the king, the names are all different. However, from David to Abraham, the geneology is exactly alike for fourteen generations of the most distant past. Matthew traces Jesus back to Solomon, the kingly line of David, while Luke traces Jesus back to Nathan the priestly line of David.

What does the Akashic Record reveal of the Solomonic Jesus? In this Jesus-child was incarnated the ego of the great initiate Zarathustra, the leader of mankind in the ancient Persian civilization. He had taught mankind of the Sun Spirit, Ahura Mazdao which his highly developed clairvoyant powers saw on the sun.

One of his leading pupils in the Persian epoch reincarnated as Hermes Trismegistus (the thrice great Hermes) who founded the Egyptian civilization. There he developed the sun-wisdom further. His other leading pupil reincarnated as Moses, leader of the Jewish people.

This great teacher Zarathustra had incarnated

again and again. In the sixth century before the Christian era, he had incarnated as Nazarathos or Zarathas teacher of Pythagoras and of the Hebraic initiates during the Babylonian captivity. (Matthew 64).

The school of initiation established by Zarathustra carried on his teachings and at the beginning of our era, the kings of wisdom of that age, saw their Leader's ego approaching earth and followed his "star" to Bethlehem. They were the three wise men, the Three Kings of the east who came to worship the Child.

Matthew then relates how this Child was taken to Egypt and remained there until Herod's death. In Egypt, the ego of Zarathustra reunited with the civilization of his pupil Hermes as well as with his other pupil who had become Moses.

This Jesus was then taken back to Nazareth. He embodied all earthly wisdom befitting the kingly line and had all the cosmic star wisdom as well.

He lived in close proximity to the Nathan Jesus who and John the Baptist were born after the massacre of the innocents. (Luke 95)

THE UNION OF THE KINGLY LINE AND PRIESTLY LINE OF DAVID

In the event of the twelve year old Jesus in the temple, a mysterious event had taken place, unsuspected and unknown to the Christian world even in our day. Rudolf Steiner's researches in the

Akashic Record reveal that the great and lofty ego of Zarathustra in the Jesus of the Solomonic line, united with the body of the Nathan Jesus when the latter was twelve years old. The pristine purity and earthly innocence of the Nathan Jesus who was irradiated with the power of Buddha was thereby filled with the star-wisdom of Zarathustra and the kingly wisdom of earth which Zarathustra had gained in all of his earthly incarnations. "Here we see the actual fusion of Buddhism and Zoroastrianism" in Jesus of Nazareth. (Luke 107). No wonder Jesus at the age of 12 surprised all the doctors in the temple, and "all that heard him were astonished at his understanding and answers." (Luke 2:47) When his parents saw him, "they were amazed . . . And they understood not the saying which he spake into them." (Luke 2:48)

The Solomonic Jesus Child died soon after the Zarathustra ego united with the body of the Nathan Jesus Child. (Could this death have thrown the Jewish esoteric leaders into confusion about their Messiah and caused their rejection of the Being in the Jesus of the Nathan line?).

Combine what is indicated above with the section on "The Mission of the Hebrew people" to follow, and one gets an inkling of the awesome divine cosmic plan to unite into one people and then into one human body the highest attainment reached by a human ego, whose physical descent included the kingly line and also the priestly line and the pure ego which came direct from the Cosmic Ego (God) and was preserved in spiritual realms to await the development of this special human body and its etheric and astral sheaths, before descending to earth.

The eighteen years from the scene in the temple with the doctors to the baptism in the Jordan by John is covered in one statement in the Bible: "And Jesus increased in wisdom and stature, and in favor with God and man." (Luke 2:52)

Rudolf Steiner's Fifth Gospel covers this period.

What does the spiritual investigator find in the Akashic Record concerning the Christ Event? When we read his discoveries for the first time, we find startlingly revealing statements and some shockingly so because our entire education in a modern world gives us only a materialistic conception of the world.

Only a few of the most important findings can be indicated here. The reader is directed to the dozen or more volumes of lectures on the subject by Rudolf Steiner.

JESUS

The Zarathustra ego lived in the Nathan Jesus from his twelfth year to his thirtieth year. During these eighteen years, he passed through the most excruciating soul experiences while traveling far and wide and visiting mystery centers. He saw how low a state the mystery wisdom of the schools of initiation had sunk. He saw clairvoyantly that demons had taken over the sanctuaries where the good gods had reigned.

His highly developed clairvoyant powers revealed the sad plight of humanity and the degradation of its spiritual leaders.

During these years, the Zarathustra ego developed the physical body, etheric body and astral body of Jesus of Nazareth to the highest degree possible.

The seven-year cycles of human development speeded up to six-year cycles in a lofty being as the Nathan Jesus and the high initiate as the Zarathustra ego. At six years, the physical body reached a certain peak of development; at twelve, the etheric body; at 18 the astral body; at 24 the sentient soul; and at 30, the intellectual soul reached a pinnacle.

The lofty being of Jesus perfected the earthly vehicles: physical body, etheric body and astral body to the highest degree that a human ego was capable of. This human ego then relinquished the three bodies at a definite moment in His life. This moment was just prior to going to John the Baptist at the Jordan.

Jesus of Nazareth had to have reached the development of the intellectual soul before the Christ Being could make use of the three lower members of his being. The Christ Being could not have made use of a child's body, even such a Child as the infant Jesus.

Therefore we see the gospels of St. Mark and

St. John omit reference to the birth of Jesus but start at the baptism in the Jordan.

THE BAPTISM IN THE JORDAN

A unique event in human and cosmic evolution took place in the Jordan when Jesus was baptized by John the Baptist. The ego of Zarathustra had already left the body of Jesus which then proceeded to John the Baptist at the Jordan.

At the baptism of Jesus, the COSMIC EGO entered the physical body, etheric body and astral body of Jesus of Nazareth, and a heavenly Voice said, "This is my Beloved Son. This day have I begotten Him." Rudolf Steiner said these are the exact words in the Akashic Record and in the earlier manuscripts. (Luke 132)

THE CHRIST BEING

Who is this Cosmic Ego? He is called the Logos by the writer of the Gospel of St. John, the Creative Word by esoteric Christianity. He is the totality of the Six Principal Spirits of the Sun, the six Spirits of Form, the six sun Elohim. (John 1908, p54)

The Cosmic Ego, the Logos had been approaching as close as the burning bush and the fire and lightning of Mount Sinai to Moses.

"The Logos was already a creating power even in the ancient Saturn Period; it then became Life while our Earth was passing through its existence as the Sun, and it became Light while the Earth was passing through the ancient Moon state." (John 1908, p43) "These Spirits of Light were at the same time the dispensing Spirits of Love." (p52)

In the earth period, during the ancient Indian civilization, the seven Holy Rishis, leaders of that epoch of culture, called this Being "Vishva Karman--the lofty Being Who was beyond their sphere." (John 1909, p11)

In the following Persian civilization, Zarathustra discerned Him as the great sun aura--Ahura Mazdao.

One of Zarathustra's two leading pupils in the Persian epoch reincarnated as Moses, the leader of the Jewish people. Moses saw that this "Spirit, Who previously had dwelt only on the sun, had moved downward to earth. He flamed up in the burning bush and shone in the fire on Sinai. He was in the elements of the earth. This Spirit proclaimed Himself as the "I AM, the I AM", i.e., the Cosmic Ego! (John 1909, p12) "Another period passed: the Spirit Whom Zarathustra had to seek in the sun, Who revealed Himself to Moses in thunder and lightning---this Spirit appeared in a human being: in Jesus of Nazareth." (John 12) This took place at the baptism of Jesus by John in the Jordan.

What must and does this Cosmic Being in Jesus say of Himself to reveal Himself to mankind as the creative Being of the solar system? He said: I AM THE LIGHT OF THE WORLD. (St. John

8:12)

This pointed to His deeds in the old Moon epoch. There He worked in the Beings Who endowed man with the astral body which gave him LIGHT---consciousness---as a soul being.

He also said: I AM THE LIFE.

This points to His deeds on old Sun, when the etheric body, the LIFE body was given to man.

THE FOUR GOSPELS ACCORDING TO ST. MATTHEW, ST. MARK, ST. LUKE AND ST. JOHN

Biblical scholars see all kinds of discrepancies in the four Gospels According to St. Matthew, Mark, Luke and John. In their minds, comparing word by word, paragraph by paragraph, or event by event, a repetition of details of each event is expected. This expectation is as justifiable as to expect four blind men examining an elephant from four points to report exactly the same details.

The seeming contradictions, when studied in the lectures on the four gospels by Rudolf Steiner, will be seen to complement each other powerfully to give mankind a fuller comprehension of the Christ Event.

Each saw this Event from his cardinal direction. As is well known, each of the gospel writers is assigned a symbol of the Zodiac: Matthew--angel; Mark--lion; Luke---bull; and John---eagle.

The writers of the gospels concentrated their testimony of Christ Jesus mainly with the events following the baptism in the Jordan. They represent the four cardinal points of the cosmos. Each therefore saw the Christ Event in a way that is unique. Each of these "seers," that is, those who have the faculty of Imaginative Cognition, emphasized his own viewpoint. In addition, John had reached a higher stage of cognition, that of Intuition as characterized earlier. He therefore revealed more than was possible with the other three.

Each of the four gospels relates events at which the writers were NOT physically present. As an example, the disciples fled before the trial and crucifixion. Yet they wrote of intimate details of those events.

The Akashic Records show that the Christ Being opened the clairvoyant powers of the writers of the gospels and they saw clairvoyantly later what their physical eyes did not and could not see when the events took place on the physical plane. They saw afterwards the spiritual traces in the Akashic Record. Of this eternal record, Rudolf Steiner says "Everything that happens in the physical sensorial world has, of course, its counterpart in the spiritual world. When a hand moves there is present not only what your eye sees as a moving hand, but behind this moving hand, this visible image of the hand, there are, for example, my thought and my will: the hand is to move. In short, a spiritual element underlies it all. But while the visible image, the sense impression of the hand motion, passes, its spiritual counterpart remains inscribed in the spiritual world and always leaves a trace; so if our spiritual eyes are opened we can trace all things that have happened in the world by the imprints left by their spiritual counterparts, Nothing can occur in the world without leaving such traces." (Gospel of St. John, Kassel, 1909, pg 16,17)

GOSPEL OF ST. MATTHEW

"The writer of the Gospel of St. Matthew emphasized Christ Jesus as Man during his sojourn on earth. All the elements of the other three gospels are present: warmth of feeling and love, knowledge full of hope and promise, majesty of the universe. It is more humanly akin to us, it is the most human of the four gospels--and describes Christ Jesus as a man. It is like a commentary of the other three gospels, clarifying what otherwise is beyond the reach of human understanding." (Matthew 22)

It starts with the geneology of Jesus tracing the blood line back to Abraham in three times fourteen generations. The importance of this direct blood line from the founder of the Hebrew people to Jesus is thereby emphasized. It points to the fact that all the qualities of Abraham's descendants were concentrated in a single Individuality, Jesus of Nazareth.

The researches in the Akashic Record by Rudolf Steiner independently confirm what is written in the four Gospels of Sts. Matthew, Mark,

Luke and John and they throw light on many passages previously unintelligible.

GOSPEL OF ST. MARK

Some of Rudolf Steiner's comments on this gospel give a clue as to how to study it:

"The Gospel of Mark, more than any other, enables us to feel the cosmic greatness and significance of the Christ." (Mark 204)

"The Gospel of Mark . . reveals the might of the creative powers permeating the universe in all their glory . . as though these forces were surging towards us from all directions of space." (Matthew 22)

Read Mark where verse after verse he describes deeds of healing, power over 'unclean spirits' who recognize and obey Jesus, and healing of many "sick of divers diseases", etc.

"In the course of the Gospel of Mark occult facts,--spiritual facts--are described side by side with purely physical facts." (Mark 188) A hint of the occult language used is the statement: "When He came to them by the sea, He sought them outside the physical body."

GOSPEL OF ST. LUKE

"The writer of the Gospel of St. Luke adhered to his purpose, which was to describe how through Christ there came into the world the infinite Love that would gradually drive out egoism," (Luke 199); and His immeasurable capacity for sacrifice. The words on the Cross resounds through the ages: "Father, forgive them: for they know not what they do." (Luke 23:34). "Words of Love, of Faith and of Hope ring out from the Cross according to the Gospel which indicates how spiritual streams that had previously been separate united in the soul of Jesus of Nazareth." (Luke 201.)

"The element of feeling speaks to our hearts, the power of His love. This gospel has been, through the ages, the source of comfort to the weak, the sick, the sinner."

The proclamation to the shepherds rightly rendered would be as follows: "The Divine Beings manifest themselves from on high, that peace may reign on the Earth below among men who are filled with good will!" (Luke 39) Compare this with the usual translation "Glory to God in the highest, and on earth peace, good will toward men." (St. Luke 2:14)

The importance of "men who are filled with good will" is the point if we are to have peace among men on earth. How sadly lacking has this good will in men been in the relationships of men in modern times!

GOSPEL OF ST. JOHN

The writer of the Gospel of St. John reached a higher stage of spiritual development than the others. He was not only a Seer, but he reached Inspiration and INTUITION as well, which made it possible for him to see deeply into the spiritual world.

He revealed the sublimity and spiritual grandeur of Christ, and points to His Cosmic nature. He starts out with the statement: "In the beginning was the Word, and the Word was with God, and the Word was God," and thirteen verses later, "And the Word was made flesh, and dwelt among us."

Human understanding is aroused by the Gospel of St. John. Our faculty of cognition is appealed to, and he reveals the magnitude of the wisdom of Christ Jesus, and the highest goal to which human wisdom can aspire and human cognition gradually attain far into the future.

THE DEED OF CHRIST AND ITS MEANING FOR WORLD EVOLUTION

The educated man of today, stripped of the knowledge of the nature of man in his higher aspects, is completely at a loss to understand such an event as the incarnation of the Logos, the Christ Being within a human body.

In future centuries, intellectual critics can even deny the existence of the gospels, or can consider them merely as superstitions prevalent up to the twentieth century. Some do so today, while other so-called Christians consider Jesus just a "good man" and some consider Him just one of the prophets.

Actually, there is NO physical proof of Jesus.

No witness recognized by the clever devil's advocate as having been present at the events in Palestine can give 'substantiated' proof of Jesus. Besides, Christianity teaches that after the crucifixion, the grave was empty and He rose into Heaven at the Ascension. Clairvoyant proof can be the only proof possible; and the gospels themselves were the result of the clairvoyant faculties given to the writers of the gospels by the power of the Christ. What they saw clairvoyantly and wrote in the gospels are confirmed by Rudolf Steiner in his researches in the Akashic Record.

Nevertheless, there will be many more among progressing mankind in the future who will have attained to Imagination, Inspiration and Intuition, and they will be able to bring additional confirmation from the Akashic Records.

THE MYSTERY OF GOLGOTHA

The Akashic Records show that the Mystery of Golgotha was the turning point in human evolution. "The aura of the earth itself changed from the moment Christ's blood flowed on Golgotha." (John 1908:112)

It is the Event from which time for progressing humanity is dated. It was a Deed for the whole of humanity, completed for every human being.

The Divine Being, Whose innate attributes encompass the solar cosmos, forsook those mighty powers in stages, in order to approach humanity and experience physical earthly life in the limited confines of a human body and to undergo physical death as a man on Golgotha.

By this deed, He incorporated into man the possibility of an upward path towards the spirit from whence he (man) had been descending since ancient Saturn.

This is the supreme earthly deed whose import is veiled in deep mystery. This is the greatest sacrifice possible in all of creation: i.e., that the Creator become as one of His created beings in order to give the latter the possibility of reuniting with the Spirit of the Cosmos.

The whole future of the solar cosmos will be needed for human egos to comprehend fully what that deed meant for humanity as well as for the cosmos and specifically for each human being.

There entered at that time into earthly evolution in Christ's deed, a Force, a Being which is a spiritual prototype for the ego, for the spirit of man. This is the Second Adam, the prototype for the spirit of man just as the first Adam was a prototype for the physical body of earth man.

If the Cosmic Ego had not come to earth, humanity would have sunk deeper and deeper into spiritual unconsciousness. The soul's diminishing consciousness after death reached an especially low point in the first centuries before the Christian era. This is emphasized by the Greek hero when he said, "It is better to be a beggar on earth than a king in the realm of the shades."

The coming of Christ brought a rejuvenating

force into earth evolution. Rudolf Steiner's researches confirms the fact that the entry of Christ into Paradise after His death on the cross, brought light among the souls of the dead. It confirms that earth evolution turned in an upward, spiritual direction since then. It depends on man's conscious thinking, feeling, and willing to be carried forward.

The momentum of the downward trend into matter, however, continues unabated into our times. Mankind must reverse the trend before it is too late.

The risen Christ became for human beings, The Example, the Prototype for each human being's future development, for all ages to come.

THE ETHERIC CHRIST AND THE SECOND COMING

Christ is now risen into the supersensible etheric realm in the circumference of the earth. He is there as the Etheric Christ, united with earth as the Spirit of the Earth. Therefore He said at the Last Supper of the bread and wine: "This is my body; . . . this is my blood." And "I am with you even until the end of earth days."

He is active there and awaits man there. Men must ascend to Him in their thinking, in their consciousness. Each one who gains for himself the faculty of Imagination has the possibility of witnessing, of becoming aware, of the 'Second Coming of Christ' even in our day.

The cases of those who have had that experience are sometimes heard of. More are probably never revealed for fear of being misunderstood. Rudolf Steiner says that the appearance of the Etheric Christ will become more and more frequent as this century approaches its end.

However, psychiatrists who know nothing of man's higher members will think that such experiences are hallucinations.

Even the one having this brilliant experience of seeing the Etheric Christ, if he does not have spiritual scientific knowledge, will miss the importance and value of the experience and it will pass him by.

Man must gain a conscious knowledge of such things. Our age requires it and demands it.

The evolution of man goes on and on from age to age to higher and higher stages. He who is satisfied with his development in any stage, no matter how high he has reached, and refuses to join the advancing stream of evolution, will be sidetracked and evolving mankind will pass him by leaving him behind.

All the pre-Christian religious leaders had been "sent ahead by the Christ to prepare mankind for His coming. He, the Logos, the Creator God has entered human evolution to lead man ever higher and higher into the spiritual cosmos. All religious streams must recognize Him and see their special development as a contribution to the whole to which they are subordinate just as the planets are subordinate to the sun.

The awakening soul can unite himself with the Christ Impulse. The prototype of human evolution whereby what one human being brings as capability into a human body can then be acquired by each human being, holds good even in the Deed of Christ: that human beings can also conquer death!

CONQUEST OF DEATH

The Mystery of Death is connected with the densest part of man, his skeleton. In a few pages, it is possible to give only the barest ideas as to the significance of the conquest of death and its connection with the skeleton. For a fuller understanding, the reader is referred to Rudolf Steiner's lectures on the Gospel of St. John, given in 1909, especially lectures 10 and 13.

The development of the skeleton took place on Earth. In the old Saturn, Sun and Moon evolution, the densest part of man had been warmth, air and water densifying to the cartilaginous condition, respectively.

Man has gained his manhood, his egohood by attaining an upright posture, due, in part to his having developed a solid skeleton. Standing on his two feet has freed his hands to carry out his impulses of will. Just contemplate a few moments what it means if man had to use his hands also for locomotion!

Man's forces, however, do not penetrate into his skeleton. He merely makes use of the gift bestowed on him. In his present evolutionary stage of development, he can change, in this life, only his astral body. That is, he can control his feelings as anger, and regulate his desires and appetites, to quite a great extent, depending on his own development. He can do much less in regard to more deep-seated qualities which reside in his etheric body, such as his temperament. In regard to his densest part, his physical body, he is completely unable to make any change by himself in this life. One might point to plastic surgery as a change, but there the work is done by another, and whenever the change is not the result of one's own inner development, it is temporal in nature.

Man's forces do not penetrate at all to his skeleton.

"At the baptism in the Jordan, the potential capacity for vanquishing death came into the world with the descent of the Christ into the threefold sheaths of Jesus of Nazareth. . . . It was a stupendous force that penetrated into the very marrow of the threefold sheath of Jesus of Nazareth at the moment of the Baptism by John". (John 151,164).

In order for evolution to advance, the form of the skeleton, as a law, has to be carried over into the future. To vanquish death requires that the laws and forces in the skeleton must be penetrated through and through by the power of an Ego, and this the Cosmic Ego, Christ, accomplished in the body of Jesus of Nazareth.

"If at the end of earth evolution he has not mastered those forces of the skeleton, he would carry over into the future nothing in the way of harvest and fruits of his sojourn on Earth had not Christ established His dominion over the skeleton."

In the evolution of the earth, "the combined powers of Lucifer and Ahriman prevented the outer world from manifesting itself in its true form. They make it appear like a world consisting only of matter and solid substance---as though no spirit underlay all matter. Instead of the mirage of the senses, man should see everywhere and in all things the divine Father principle, of which all things and he himself are a part. Man should see in all the sense world around him, their origin is in spirit: spirit underlies our physical-sensible world. The primordial form of spirit from which springs all that is physical and of the senses is the aspect of divinity, known in Christian esotericism as the Father principle. This had to be veiled by Ahriman for a time so that man can gain freedom.

The veils cast over the sense world, obscuring the Father principle became densest in regard to death. "Since death is interwoven in the totality of the sense world, it is something that pertains to the divine-spiritual Father principle . . . . What must man seek? The Father, the Cosmic Father; and just as he must learn to think of every object as being in truth the Father, so he must come to feel that death, too is the Father." (John 204).

The false picture of the Father principle, "the grotesque image appearing so deceptively as death is caused by the Lucifer Ahriman principle infused into every phase of our life."

To disabuse man of the false, deceptive view of death, a Being having nothing to do with any cause of death among men, a Being in no way connected with any cause of human death, that is, with anything brought about by Lucifer and later by Ahriman, or with any deeds done under the Lucifer - Ahriman influence--in short, with anything whereby men became guilty or fell a prey to evil--- only an utterly undeserved death, undertaken by one without guilt--- could extinguish all guilty death.

An innocent Being, accordingly, had to suffer death, submit to death; and by so doing He infused into human life those forces which will gradually create knowledge concerning the true aspect of death; that is, the realization that death as it appears in the sense world is not truth---that on the contrary, this death had to occur to provide for life in the spiritual world.

Thus the innocent death on Golgotha furnished the proof which will gradually be comprehended by humanity, that "DEATH IS THE EVER-LIVING FATHER". (John 206).

CHAPTER VI

The Mission of the Hebrew People

A study of Rudolf Steiner's lectures on the four gospels of Sts. Matthew, Mark, Luke and John impresses the reader with the uniqueness of the mission of the Hebrew people up to the time of Christ, in furthering earthly evolution.

They were THE CHOSEN PEOPLE among all those on earth at that time for that particular mission and task in advancing human development. They were literally chosen to bring man's ego more fully into his physical body.

"Abraham is spoken of as the inventor of arithmetic. He is therefore to be regarded as the first personality into whom was implanted the physical organ of thinking, the organ through which man, by means of physical thinking, could rise to actual thought of the Divine, whereas formerly it was only through clairvoyant vision that he could have any knowledge of God and the Divine." (Matthew 67)

The complex Divine plan to prevent humanity from plunging too deeply into earth-life and thereby becoming diverted from his cosmic goal is not easy for the human mind to understand, especially the mind that wants everything to be simple.

In the lectures on the Gospels, Rudolf Steiner traces the coming together of the main streams of human development into the Hebrew people.

In Moses, there reincarnated one of the two leading pupils of the Original Zarathustra, founder of the old Persian civilization. Zarathustra had given Moses his etheric body which included all the wisdom of Zarathustra, who saw the Sun Being Who he knew would one day approach earth. Moses saw this Sun Being approach in the elements of earth in lightning and fire on Sinai, and in the burning bush, but Moses did not recognize Him as the Christ Being, so he was not permitted to enter the promised land. "He was chosen to be the means whereby the great primeval wisdom of the past could be carried over to the time of the Mystery of Golgotha" (Mark 164,165). The initiation-secrets of the whole of the rest of the world passed into the Jewish people through Moses through his meeting with the Midianitish priest, Jethro."

The other leading pupil of Zarathustra was reincarnated as Hermes to whom was given the astral body of Zarathustra. Hermes was the founder of the Egyptian civilization.

The Hebrew people as a whole encountered the Zoroastrian stream as it developed further as the reincarnated Zarathustra taught the Chaldean and also the Hebraic Initiates as Nazarathos in their Babylonian captivity, which was spent just across the Tigris-Euphrates Rivers from ancient Persia. They also encountered the Hermetic stream in their sojourn in Egypt.

Why was it necessary that there be a uniting of the many streams of human development? Each stream of culture developed some definite human capacities to an especially high level. The leaders of each culture were the high Initiates who guided the development of each people. The leading streams of human development had to be united in one people, and eventually in one Individuality.

The "Hebrew Prophets are the reincarnations of the Initiates of other peoples and had passed through certain stages of initiation in other lands . . . Jeremiah, Isaiah, David and so on are to be found among other peoples when we trace their souls back into earlier incarnations." (Mark 37)

"In them was embodied all the ancient wisdom of the Occult Mystery Schools of the pre-Christian world. These spiritual leaders entered Jewish history like brilliant comets. They came, seemingly, out of nowhere and burst upon the historical scene as the Prophets."

"What formerly had been scattered was now to be gathered together into one focal point, and was to be born again from out of the blood of the people of the Old Testament." (Mark 38)

Forty-two generations were needed to bring in the descendants of Abraham through the Solomonic line (Gospel of St. Matthew), the development and perfection required where the intellectual soul was concerned. The other stream flowed through the Nathan line of David (Luke Gospel). Then it was ready for the entry of the Cosmic Ego into the physical body of Jesus of Nazareth.

The whole mission of the Jewish people was centered in this preparation.

This culminated in the birth of Jesus of the Nathan line of the house of David as specifically given in the Gospel of St. Luke; and of the Jesus of the Solomonic line of the house of David as given in the Gospel of St. Matthew. Both sets of parents were named Joseph and Mary! Coincidence? Or Cosmic Plan?

Note how Jesus was named. To Mary in St. Luke's Gospel "Thou shalt conceive in thy womb and bring forth a son, and shall call his name JESUS." (St. Luke 1:31)

To Joseph: "Thou shalt call his name JESUS." (St. Matthew 1:21). Both of these were the directions of "the angel of the Lord." (Matthew 1:20; Luke 1:30) Why these two streams were needed and how they meet is touched on under the subject of JESUS.

Since natural clairvoyance had become dimmed to a large extent especially among the Jewish people who were to develop their own ego force, the presence of the Cosmic Ego in Jesus Christ could not be recognized by the decadent ruling Jews of Jesus' time.

How decadent were they? We read in Matthew 23, Jesus' indictment: "The scribes and Pharisees sit in Moses' seat: All therefore whatsoever they bid you observe, that observe and do; but do not ye after their works: for they say and do not; . . . they bind heavy burdens and lay them on men's shoulders; but they themselves will not move them with one of their fingers."

There follow 36 verses of indictments, twelve of those verses beginning with "Woe unto you, scribes and Pharisees, hypocrites! . . . ye blind guides . . . ye fools and blind" each with a charge: "Ye shut up the kingdom of heaven against men . . . ye devour widows' houses . . . for a pretense make long prayer . . . make one proselyte and when he is made, ye make him twofold more the child of hell than yourselves . . . ye have omitted the weightier matters of the law, judgment, mercy and faith (ye are) full of extortion and excess . . . full of hypocrisy and iniquity."

"Ye serpents, ye generation of vipers, how can ye escape the damnation of hell? . . . . . Jerusalem, thou that killest the prophets, and stonest them that are sent unto thee." (St. Matthew 23)

Such were the old decadent establishment leaders of the Jewish culture of the time of Christ Jesus, who condemned and had Him crucified.

Those scribes and Pharisees were condemned

by Jesus Christ as "hypocrites, blind guides, fools, serpents, vipers," yet their decision of not accepting Jesus Christ as the Messiah and their non-recognition of Him, has been perpetuated and permitted to stand by almost twenty centuries of Jewish leaders. This has resulted to the detriment of at least sixty generations of their descendants, and the end is not in sight.

Those of the present day have the benefit of almost two thousand years of history to judge the power of the Christ Impulse, as it weaves and works in the highest examples of Christian living and civilization; or they can study the teachings of the New Testament to see whether they fulfill the prophecies of the Old Testament.

But they cling tenaciously to what is old and refuse to join the advancing stream of humanity which their race had made possible. In truth, they continue to refuse to accept credit for the greatest event in human history--the incarnation of God in humanity.

"The Jewish people ought to have known that a conclusion, and a RENEWAL, of their own peculiar mission was taking place. The Jews should have understood that He represented a climax to their own history." (Mark 181)

However, humanity owes to those other Jews now known as St. Peter, St. John, St. Matthew, Mark, Luke, James and the other Disciples as well as Mary, the mother of Jesus, their recognition of

Jesus Christ.

Mankind has entered into the age of the spiritual soul in our day, so men of goodwill of all nations, if they WILL, can see the Christ Impulse at work among all men: the Red Cross, the philanthropy extending beyond color, race, and national boundaries, the growing feeling of brotherhood and the concern for people everywhere. These impulses become strongest in times of natural disasters as floods and earthquakes, or when earthmen are imperilled in space.

Since human institutions became stagnated and are difficult to change, it is the individual's own initiative and responsibility to recognize and accept the truth which his own sense of truth gives him.

By their own free WILL, our present generation and those to come can unshackle themselves from the imposition of the disbelief of long-dead rabbis who abused, tortured and crucified the Creator God who lived as the Son of Man for three years, lived and repelled the Luciferic and Ahrimanic tempters of man and who stood before humanity as the Ideal, the Prototype of what he can become at the end of earth evolution. But they spat on Him, scourged Him, and brought Him to the place of crucifixion.

To escape from the narrow confines of the decisions of a few leading conservative decadent rabbis of twenty centuries ago in denying that Christ had come in such humble surroundings, is the responsibility of present-day Jewish leaders. They have the advantage of a historical survey of two millenia, over those of Christ's time.

It will be to their everlasting credit to acknowledge the fulfillment of the mission of the Hebrew people for the whole of humanity. In making the Christ Event possible, the Hebrew people performed a service for every human being, incarnate as well as discarnate.

If human beings of today retained the belief of 2000 years ago that the earth is flat, we would consider him ignorant or worse. To reject the Christ Impulse as it has worked during the last nineteen and a half centuries is of the same order of intelligence.

The responsibility ultimately rests on the individual alone. Each ego must eventually make his own decision. His own evolution depends on his own thinking, his own feeling and his own willing.

CHAPTER VII

Lucifer and Ahriman

According to the College Standard Dictionary, Lucifer is "Satan, the prince of darkness: so called by the church fathers, under the impression that the passage (Isa.xiv,12) 'How art thou fallen from heaven, O Lucifer, son of the morning!' had reference to him, instead of to the king of Babylon."

"Ahriman", notes the same source: "In Persian mythology, the Prince of Darkness; evil deity; the author of evil and ruler over the kingdom of darkness. In the later Persian dualism, he held equal sway over the world with Ormuzd, the good deity."

"Satan" according to Webster's Unabridged Dictionary, "(Heb. satan, enemy, from satan, to be adverse). The great adversary of man, the devil or prince of darkness, the chief of the fallen angels."

"Devil. In theology, an evil spirit or being, specifically the evil one, represented in Scripture as the traducer, father of lies, tempter, etc." Webster's Unabridged.

"Ahriman, Ariman: (Per). The evil genius or demon according to the dualistic doctrine of Zoroaster. Ariman is the personification of malignity, the original source of all moral and physical evil, the chief of the devils and malignant spirits, the king of darkness and of death, and the foe of Ormazd and his kingdom of light and life." Webster's Unabridged.

These definitions refer to ancient beliefs---early church fathers or the ancient Persians.

What does modern man think about Lucifer and Ahriman? Are there such beings or are they fanciful superstitions of non-scientific cultures?

If the question is put to the citizens of this country, there would be those who believe strongly that there is such a being, those who feel just as strongly that there are no such beings, and some in between who feel that they have no opinion. Those who believe there is, would think Lucifer and Ahriman are one and the same being.

A well-known theologian said he does not consider it rational to hypothesize a being who wills evil, so he apparently dismisses these beings as fictional and non-existent. The theologian may not be able to conceive of such a being, but unfortunately, the existence of such beings does not depend on what the theologian can conceive.

Most of the inventions of the world could not be conceived of by the forefathers of our philosophers and theologians too, but nevertheless the inventions we use daily as the airplane, radio and TV became realities in spite of the doubts of 'authorities'.

Christians could give many such examples as 'Jesus casting out devils', and the three temptations of Christ Jesus 'by the devil' after the baptism in the Jordan. But how many Christians consider such statements as figures of speech or allegory which do not refer to real beings? Most people will think: "This is a scientific age. All talk of the devil or satan is nonsense. They have nothing to do with me."

However, one of the greats of the human race wrote a whole book about the activities and deeds of a being named Mephistopheles and of how he tempted a man as modern as any scientist of our age. Goethe had a deeper insight into Reality than modern man is willing to admit. Even if they concede greatness to his Faust, they do not think or believe he was describing a real being.

What does a modern Initiate who has developed higher faculties of perception say about Lucifer and Ahriman?

Before the separation of the Earth from the sun, "the Luciferic beings did not advance as far as the moon gods. They could not exercise their power in the fire mist. But when man developed the power for human understanding, these beings seize the opportunity to direct man's understanding towards the things of earth. They gave man the predisposition to freedom, to the discrimination between 'good'' and 'evil'.'' (Cosmic Memory 138-140)

In ancient Saturn, the substance of the Thrones laid the first foundation of the human physical body. In old Sun, the Spirits of Wisdom imbued men with the life-body. On old Moon, the soul was added by the Spirits of Motion. On Earth the sentient soul was added.

"Lucifer anchored himself in the sentient soul, and there he remains. On Earth the Spirits of Form endowed man with the I or Ego. In Lemuria the Luciferic beings set themselves in opposition to the Spirits of Form. This gave man the prospect of freedom; also the possibility of evil-doing, of succumbing to desire and passion in the world of sense.

"The Luciferic beings took hold of man's astral body, thereby debasing his passions, urges, desires. They made men have the interest, and to crave the impression of the earth."

There you have in a nutshell, THE ORIGIN OF EVIL.

"The Spirits of Form took countermeasures: they inculcated into the being of man illness, suffering and pain. This was the necessary counterweight to the deeds of the Luciferic spirits." (Deeds of Christ and the Opposing Powers 10)

There one must also see that debasing his passions, urges and desires are the ultimate causes of illness, suffering and pain. These latter serve to give man the power to counteract and overcome his debased passions, urges and desires. Illness, pain and suffering therefore, taken in the right way, strengthen the human being and give him power to rise to a higher level. They must not be taken as a curse of God, but as a gift to save the soul from the clutches of Lucifer whose aim is to lead him astray from earth evolution.

THE INCARNATION OF LUCIFER

In the third millenium before the incarnation of Christ in the flesh, there was an incarnation of Lucifer in the flesh in the east. It was specified to be in China. Much of the wisdom of the mystery centers of the 'pagan' world stems from this Lucifer incarnation.

"About the beginning of the third millenium before the Mystery of Golgotha a momentous event took place. Before this event, man was still unable to make use of the organs of intellect. These organs were already within him, they had taken shape in his physical constitution, but the being of spirit-and-soul could not make use of them. Thus men could not acquire knowledge through their own thinking, through their own powers of intellectual discernment. They were dependent upon what was imparted to them from the mysteries. And then, about the beginning of the third millenium, a momentous event took place in the east of Asia.

"A child of a distinguished Asiatic family was allowed to grow up in the precincts of the Mystery ceremonies . . . . was permitted to take part in the ceremonies . . . . this man, suddenly at the age of about 40 began to grasp through the faculty of the human intellect itself what had formerly come into the Mysteries through revelation and only through revelation. In this man Lucifer himself was incarnated---no more and no less than that. And from this (Chinese) incarnation of Lucifer in the flesh, there went forth what is described as the pre-Christian, Pagan culture which still survived in the Gnosis of the earliest Christian centuries." (Lucifer and Ahriman 37)

"The Luciferic beings brought into earth evolution the world of thought, of intellectualistic thought pervading culture, the world of reason. Lucifer's intention was to convey the wisdom to men in such a way that it would induce them to abandon the path of earth-evolution and take a path leading to a super-earthly sphere, a sphere aloof from the earth. Luciferic beings aimed to make men turn away from earth, to win men for a kingdom alien to the kingdom of Christ. But the wise men of olden days pledged themselves to use the wisdom for the good of earth-evolution." (L & A 56)

The Luciferic beings aim to make men premature spiritual beings of a low order. They wish to make men avoid earth-life and thereby miss the opportunity to gain freedom which is one of the purposes of earth evolution.

AHRIMAN

"Ahriman (Satan) from the middle of the Atlantean period, enticed man to regard everything in his environment as material, thereby hiding its true spiritual foundations in stone, plant, and animal. Ahriman established his footing in man's intellectual soul, the transformed part of the etheric body. From there he lures man to false conceptions and judgments of material things, leads him to error, to sin, to lying. In every illusion that matter is the sole reality, we must perceive the whispered promptings of Ahriman."

"As a countermeasure to the workings of Ahriman, the SPIRITS OF PROGRESS introduced Karma. Karma is the gift of Christ; it is a power for which man should be thankful. Without Karma, no progress would be possible. An error demands that one shall right it, then having expunged it from one's karma, one can again take a step forward." (Deeds of Christ and the Opposing Powers)

The Ahrimanic beings have an opposite aim to Lucifer. They strive to make men earthbound. They harden men, make them sclerotic and dried up.

"In earthly life, the power of Ahriman misleads man into regarding the sense-perceptible, physical existence as the one and only reality, thus shutting himself off entirely from any kind of outlook into a spiritual world. In the spiritual world, Ahriman brings man to complete isolation, leading him to center all his interests upon himself alone. Men who at death are in the power of Ahriman are born again as egotists." (Occult Science 213)

Just as there was an incarnation of Lucifer in the third millenium before Christ, so will there necessarily be an incarnation of Ahriman in a human form in the third millenium after Christ.

"What is now in preparation and will definitely come to pass on earth in a none too distant future, is an actual incarnation of Ahriman."

Only the priests in the mystery temple in China were aware of the incarnation of Lucifer, and men were very unconscious of what the incarnation of Christ and the Event of Golgotha signified "BUT it is of the utmost importance that men should approach the coming incarnation of Ahriman with full consciousness of the event." (Lucifer and Ahriman 38)

Ahriman prepares for his incarnation in a number of ways. This incarnation will take place in the West. "A Being like Ahriman, who will incarnate in the West in time to come, prepares for this incarnation in advance . . . Ahriman guides certain forces in evolution in such a way that they may be of greatest possible advantage to him. And evil would result were men to live on in a state of drowsy unawareness, unable to recognize certain phenomena in life as preparations for Ahriman's incarnation in the flesh."

" . . . . the time has come for individual men to know what tendencies and events around them are machinations of Ahriman, helping him to prepare for his approaching incarnation . . . . It would undoubtedly be of the greatest benefit to Ahriman if he could succeed in preventing the vast majority of men from perceiving what would make for their true well-being, if the vast majority of men were to regard these preparations for the Ahriman-incarnation as progressive and good for evolution. If Ahriman were able to slink into a humanity unaware of his coming, that would gladden him most of all. (He has already succeeded to a very great extent). It is for this reason that the occurrences and trends in which Ahriman is working for his future must be brought to light."

"One of the developments is the spread of the belief that the mechanistic, mathematical conceptions inaugurated by Galileo, Copernicus and others, explain what is happening in the cosmos . . .

"It would augur success for Ahriman's temptings if men were to persist in merely calculating the revolutions of the heavenly bodies, in studying astro-physics for the sole purpose of ascertaining the material composition of the planets---an achievement of which the modern world is so proud. BUT WOE BETIDE IF THIS COPERNI-

CANISM IS NOT CONFRONTED BY THE KNOWLEDGE THAT THE COSMOS IS PERMEATED BY SOUL AND SPIRIT. It is this knowledge that Ahriman, in preparing his earthly incarnation wants to withhold from men."

Men in the last few centuries have become so intrigued with life on earth that a counter-force must be created to combat the Ahrimanic impulse which would drag men down into a culture devoid of spirit. Mankind has swung from one side of the pendulum to the other.

It is imperative that mankind keep the balance between the two powers, Lucifer and Ahriman, while on earth. The only force possible is the modern initiation wisdom which a science of the spirit gives.

This is the body of wisdom including the entire life-work of Rudolf Steiner, comprising about 6000 lectures to which he gave the name ANTHROPOSOPHY, i.e., anthropo, man; sophia, wisdom---that can be characterized in numerous ways, i.e., the wisdom that 'unites the spirit of man to the Spirit of the Universe', the wisdom of the cosmos that is embodied in the being of man.

Individuals, in life, swing from Ahriman to Lucifer, the one pushing him into the hands of the other. Ahriman hides the true spiritual foundation of stone, plant and animal, so man considers all the food he eats as devoid of spirit. Therefore when man devours his food, he consigns them to the realm of Lucifer.

Only when he holds in his consciousness that Christ's words "This is my body, this is my blood" apply to all his food also, will he redeem the food he eats to the realm of Christ. He who does this will be more likely to consider his own body as a holy vessel to be used wisely and to dedicate his life efforts to the advancement of man.

Humanity is deeply engulfed in the study of the physical aspects of the universe. This is the aim of Ahriman. Many leading minds consider this mechanistic pursuit a noble and progressive development, but have no way to know that this is Ahriman's plan, because they do not know that Ahriman exists. Most professors of science no doubt consider anyone who speaks of Ahriman to be a lunatic. They do not know that they themselves are the dupes of the cleverest seducer in the solar cosmos, although a more powerful one will enter human evolution more and more as the twenty-first century approaches.

Think of the billions and billions of dollars and rubles expended to bring back samples from the moon and the continuing expenditure of endless billions of dollars and rubles to take physical measurements of the most sophisticated kinds in order to determine whether conditions suitable for life exist on the planets.

If a few pounds of material from Mars or some other planet should be brought back to earth, especially if a leaf of a fern or some microorganisms were included in the samples, it would be considered a major achievement of the 1970's and all the untold billions would be considered well-spent.

Such an achievement would be hailed by press, radio, TV and the whole scientific world as a prelude to the solution of the mystery of life. With all the teeming life on earth, scientists are at a blind alley in understanding life, since they will not consider the etheric, soul, and spirit world a reality. Life ether streams through the cosmos: LIFE PERMEATES THE COSMOS!

In the inevitable event that a small sample of extraterrestrial life is found, it will be hailed as a major break-through toward the solution of life in the universe.

This will be mere wishful thinking. If scientists cannot understand the teeming life on earth, neither will they understand life when it is found in the cosmos.

Such an achievement of discovering living specimen on another planet would add as much to man's understanding of the universe as a microbe-geologist gains of man by chipping away at a finger nail of a sleeping archeologist. He might miss and make an astouding discovery: there is a crimson flood pouring from this specimen. The microbe-geologist is lucky to escape with his life. He rushes back to his laboratory and analyzes his chippings and a part of the flood. The microbe-scientists are astounded; they organize expeditions to bring back additional samples. Microbe-businessmen organize caravans to bring back what may be a precious substance---that red flood. But when they reach the spot, the big hunk had disappeared. So the mystery remains as mysterious as ever, and as insoluble.

No, Mars and Jupiter will not disappear, but the Spiritual Beings who have their center of activities on Mars and Jupiter will not be approachable by the tools of our scientists, and their tremendous efforts will be in vain. Life cannot be solved by physical scientific means even if all the universities on earth think it can be done. They will get no farther than the microbe-geologist can in understanding man.

Ahriman tempts man to have a repugnance for knowledge concerning soul and spirit in the cosmos. The mineral, plant and animal worlds are considered to be devoid of soul and spirit. If one were to hint that elemental beings may be active within these kingdoms, the modern intellect will consider such ideas a relic of paganism.

However, if man can comprehend that matter is a condensation of spirit and that before there was a gaseous nebula, there were already the Spiritual Hierarchies, the kingdoms of nature will take on a more illuminating aspect. Then mankind will realize that there is more truth than they now suspect in the fairy tale of the unconscious Sleeping Beauty and the prince: the consciousness of the prince that Sleeping Beauty was a soul being and his love for her were the forces which revived her and released her from the spell of unconsciousness.

Our recognition, our consciousness of the kingdoms of nature as condensations of spirit, originating from spirit, and taking the form they do so that man can incarnate as earthmen, will release even the food that we eat from the spell in which Ahriman wishes to keep the soul beings connected with the kingdoms of earth. This does not mean a pantheistic idea of an abstract spirit pervading all of nature, but a specific understanding of the elemental beings.

Or we could take seriously the statement of Christ at the Last Supper when He said of the bread "This is my body," and of the wine, "This is my blood." Christ points to the fact that the whole earth is His body and the fluids on it form His blood, because He united with the Earth and is the Spirit of the Earth since the Mystery of Golgotha. "I am with you alway, even unto the end of Earth days."

Astronomers will keep building visual and radio telescopes of higher and higher power in many places on earth. They will take as many photographs and measurements as possible. Eclipses are photographed from jet planes and distant mountain tops. The devotion in gathering physical data shows man's intense desire to know more about the worlds around him, but the solution of those mysteries of life and of the cosmos can never come by physical measurements, because matter is a condensation of spirit and to deny the existence of spirit closes the door to its understanding.

Radio telescopes covering acres monitor incoming signals from deep space in order to detect any possible messages from intelligent beings in other systems who may be trying to communicate with earth.

If man would devote himself to gaining an understanding of those higher beings who work on his thinking, his feeling, his willing and the opposing powers who strive to divert him from the evolutionary course planned by his creators, he would be living more in a way to forward his evolution.

The second of Ahriman's endeavors "is to preserve the already wide-spread attitude that for the public welfare it is sufficient if the economic and material needs of men are provided for." (Lucifer & Ahriman)

Education is centered in getting training in the professions to gain the most prestige or to get the best jobs. The sciences give an understanding of external nature, but contribute nothing to an understanding of the soul and spirit of man. In fact, modern man is not sure whether he has a soul and spirit. Conversations leading to discussions of these subjects will usually draw a blank stare and a fast retreat. The average man seems uncomfortable, has a fear, talking about such an 'unlikely' thing as soul and spirit. He probably considers it a superstition.

That is exactly the attitude Ahriman seeks to develop in mankind, and anyone familiar with our universities knows that this seducer of man has succeeded to an unparalleled extent.

Material needs are considered so desirable that some men will do anything to wrest as much as possible from others, no matter how much other human beings suffer from their methods. There are crimes of all sorts, organized and unorganized. Widespread theft in all levels of society, from the public in streets and homes, from stores, employers; by officials in government; stolen automobiles that number as many as a million in a year. Another is the selfish tendency of getting as much as possible without giving anything in return even if the recipient is able to contribute something for what is given him. Our proliferating welfare and medicare problems are indications of what this attitude will lead to. Another is the widespread tendency of groups which make exorbitant demands resulting in hardship to the general public.

"The Ahrimanic incarnation will be greatly furthered if men fail to establish a free and independent spiritual life and allow it to remain entangled in the economic or political life."

(Lucifer and Ahriman 22) The free and independent spiritual life of man consists in each person having the freedom to choose not only his religion, but his own education and school, and the direction he wishes to develop his talents.

"If man is not alert and remains in drowsy insensibility and permits to pass unnoticed the part relating to religious freedom, he will miss the repetition of the Mystery of Golgotha upon the etheric plane, the reappearance of the Etheric Christ . . . . This is the great danger for the period of the consciousness soul. That is what might occur if men should not turn to the spiritual life before the beginning of the third millenium, before the year 2000."

The activities in the brain of man are studied and traced to electrochemical processes. The activities of ego, astral body and etheric body are not known. Memory is likened to storage of data in computers. The part played by the etheric body in memory is not suspected. It is the hope of such researchers that once they can trace all the connections that are in the brain, they will know all there is to know of man's power of thinking. Love and devotion will no doubt be attributed to one or another hormone, acting illogically, and a Goethe or a Shakespeare to a chance combination of genes. The past lives and connections with living spiritual inspirers are not suspected, much less are they known.

The third particular advantage for Ahriman's

incarnation "is the short-sighted, narrow conception of the Gospel that is so prevalent today. Men wish to take the Gospels 'in all their simplicity', just as they stand." (Lucifer & Ahriman 14)

In the first place, "we no longer possess the true Gospels. The translations are not faithful reproductions of the authentic Gospels." (See section on several examples frequently quoted). If men base all that they stand for on the errors in translations, such narrow-minded conceptions can lead to folly.

In the deeper aspect, "no true understanding of Christ can be reached by the simple, easy-going perusal of the Gospels beloved by most religious denominations and sects today." (Lucifer & A.15) In the first centuries after the Mystery of Golgotha, men still had the Luciferic wisdom that had been handed down by tradition, but in our day such wisdom has been lost.

Men therefore will not be able to reach the real Christ for Whom spiritual science seeks. As an example, how many Christian sects look forward to the Second Coming of Christ in a physical body, and thus coming again into the realm which Ahriman dominates?

Christians of many sects pay no attention to Christ's words: "MY KINGDOM IS NOT OF THIS EARTH." Instead of stepping upward to the next higher realm to follow the risen Christ up into the spiritual spheres, so-called Christians expect Christ to become an earthly king just as Judas Iscariot did. An earthly kingdom is what Ahriman strives to foster and create for a misled, deluded humanity.

"That Being Whom we call the Christ appeared on earth in the flesh at the beginning of our era. He will never come again in a physical body, for that event was unique. But the Christ will come again in an etheric form. Then men will learn to perceive the Christ, because through this etheric vision they will grow upward toward Him Who now no longer descends so far as into a physical body, but only as far as an etheric body. Therefore it will be necessary for men to grow upward to a perception of Christ. For Christ spoke truly when He said, "I am with you always, even unto the end of the Earth cycles." He is here; He is in our spiritual world . . . . And those who are especially blessed can perceive Him always in this spiritual-etheric world.

"It was Paul, in particular, who was convinced through such perception, in the event of Damascus. But this etheric vision will be developed as a natural faculty by individual persons to experience an event of Damascus; a Paul-event, will be more and more possible for men in the coming period (1930's onward)." (The Vision of Christ's Advent in the Etheric January 25, 1910, pg.13)

"It will then no longer be necessary to prove

the existence of Christ through all sorts of documents, for there will be eye-witnesses to the presence of the living Christ in those who will know Him in His etheric body.

"The greatest mystery of our time is this one concerning the second coming of Christ, and thus it appears in its true form." (Vision of Christ's Advent in the Etheric 14)

Modern theologians speak of 'the simple man of Nazareth', as if Christ were only a little more exalted than prophets, saints and founders of other religions. They also identify Christ with Jesus of Nazareth. This 'simple man of Nazareth' idea leads only to an hallucination, to a kind of illusion." An illusory conception of Christ is all that can be gleaned through the way in which the Gospels are read today---not to the reality of Christ." (Lucifer & Ahriman 15).

"Sects which swear by the literal content of one Gospel alone or another is an illusory conception stemming from a dimming of consciousness. With the dimming of consciousness that inevitably occurs when the deeper content of the Gospels is not revealed, men would fall wholly into Ahriman's service, helping in the most effective way to prepare for his incarnation, and adopting towards him the very attitude he desires."

The point to remember is, however, that the people who do most to prepare for the incarnation of Ahriman are those who constantly preach:

"All that is required is to read the Gospels word for word---nothing more than that!" (Lucifer & Ahriman 16)

"Nationalism and all that is inherent in it is another development favoring Ahriman's incarnation. Whatever can separate men into groups, whatever can alienate them from mutual understanding the whole world over and drives wedges between them strengthens Ahriman's impulse. His interests are promoted too by the fact that men are taken up with the most divergent shades of party opinions, of which the one can be justified as easily as the other." Hatred and bitterness will be intensified. In reality we should recognize the voice of Ahriman in what is so often proclaimed nowadays as a new ideal: 'Freedom of the peoples, even the smallest', and so forth. Blood relationship has ceased to be the decisive factor and if this outworn notion persists, we shall be playing straight into the hands of Ahriman!" (Lucifer & Ahriman, 1919, 14) These words were spoken in 1919. How far along Ahriman's road we have travelled in the formation of new nations since then! Unions, parties of all shades of opinions, floods of articles printed daily aimed at dividing men, all work on behalf of Ahriman's aims to split humanity into smaller and smaller factions.

In Ahriman's incarnation, he will "establish a great occult school for the practice of magic arts of the greatest grandeur, and what otherwise can be acquired only by strenuous effort would be poured over mankind . . . . Lovers of ease who refuse to have anything to do with spiritual science would fall prey to his magic, for by means of these stupendous magic arts he would be able to make great numbers of human beings into seers--but in such a way that each individual would be strictly differentiated. What one person would see, a second and a third would not see. Confusion would prevail . . . and men would inevitably fall into strife on account of the diversity of their visions. The result would be the establishment of Ahriman's kingdom on earth and the overthrow of everything hitherto achieved by human culture.

"If we are to acquire insight into the nature and evolution of man, we must be constantly mindful of the power and influence of Lucifer, of Christ and of Ahriman." "Ahriman's plans are to strive to kill in man the consciousness of his own spirituality, to induce in him the belief that he is only a completely developed ANIMAL (Ahriman is in reality the great teacher of materialistic Darwinism); to teach all technical, practical activity that admits the value of nothing beyond the external life of the senses; to produce an extensive technology only so that man may satisfy with greater finesse, hunger, thirst, and other animal needs; and to kill in man the realization that he is an image of Deity.

Man must become alert to the brilliant work

of Ahriman as an author through the whole of the twentieth century. "Even in our day, much in the subconsciousness is educating souls in such a way that they will be able to incarnate again quickly and become instruments for Ahriman as an author. He will write in all domains: in philosophy, in poetry, in the sphere of the drama and the epic; in medicine, law, sociology; Ahriman will write in all domains. The Ahrimanic character must be recognized."

What can man do to keep his equipoise and even to overcome the onslaughts of the Luciferic and Ahrimanic forces which constantly confront him in life?

"THERE IS BUT ONE POWER BEFORE WHICH LUCIFER RETREATS, AND THAT IS MORALITY WHICH BURNS HIM LIKE THE MOST DREADFUL OF FIRES."

"AND THERE IS NO MEANS BY WHICH TO OPPOSE AHRIMAN OTHER THAN A POWER OF JUDGMENT AND DISCERNMENT SCHOOLED BY SPIRITUAL SCIENCE. FOR AHRIMAN FLEES IN TERROR FROM THE WHOLESOME POWER OF JUDGMENT ACQUIRED UPON EARTH. IN THE MAIN THERE IS NOTHING FROM WHICH HE HAS GREATER AVERSION THAN THE QUALITIES WE GAIN FROM A HEALTHY EDUCATION OF OUR EGO CONSCIOUSNESS. FOR WE SHALL SEE THAT AHRIMAN BELONGS TO A VERY DIFFERENT REGION FAR REMOVED FROM

THAT FORCE OF SOUND JUDGMENT WHICH WE DEVELOP IN OURSELVES.

"THE MOMENT AHRIMAN ENCOUNTERS THIS, HE RECEIVES A TERRIBLE SHOCK, FOR THIS IS SOMETHING COMPLETELY UNKNOWN TO HIM, AND HE FEARS IT. THE MORE WE APPLY OURSELVES IN OUR LIFE TO DEVELOP THIS WHOLESOME JUDGMENT, THE MORE DO WE WORK IN OPPOSITION TO AHRIMAN." (Manifestations of Karma 136)

CHAPTER VIII

Sleep

Sleep is the subject of intensive study by scientists. Using instruments to detect rapid eye movement (REM, the term used by researchers) electrodes are taped to various parts of the body to take measurements of changes in the activity of eye, heart and other organs.

What happens to the astral body and ego during sleep will remain a mystery to scientists who do not recognize the existence of these members of man's being. Quarter-truths will result from these studies.

In sleep, pain, pleasure, joy, grief, thoughts, mental images, and the power of giving conscious direction to the will are blotted out. Of all these, the astral body is the bearer. In sleep, the astral body vanishes from external observation. Supersensible perception must now trace it through the stages of its life till awakening when it once more takes possession of the physical and etheric bodies. (Occult Science, Chapter 3).

Having been active within these bodies in waking life, the astral body must again and again in rhythmic succession enter into the pure world of soul and spirit. This need is felt by the soul and spirit as tiredness and fatigue.

"Of all that the Spirit has to accomplish on earth through man, by far the greater part is accomplished during sleep. While we are asleep, lofty Spiritual Beings work upon the human soul, with the object of bringing man to his full and complete evolution in Earth existence." (Man's Life On Earth and In the Spiritual Worlds 42).

In sleep, man's astral body and Ego are outside his physical and etheric bodies. "But it must not be imagined that astral body and ego during sleep are like a misty cloud hovering in the vicinity of the physical body. What an inferior kind of astral clairvoyance sees in the form of a cloud and which we call the astral body, is only the crudest beginning of what the human being reveals during sleep." Anyone regarding this cloudlike formation near the physical and etheric bodies as the sole phenomenon of importance would simply be basing himself upon the lowest form of astral clairvoyance.

"The truth is that during sleep, man is a being of vast magnitude. At the moment of going to sleep, the inner forces in the astral body and in the Ego actually begin to expand over the whole solar system, to become a part of it. From every direction man draws into his astral body and into his Ego, forces which strengthen his life during sleep, and on waking he contracts into the narrower confines within his skin and pours into these what he has absorbed during the night from the whole solar system. That is why medieval occultists too called this spiritual body of man the 'astral body', because it is united with the world of the stars and draws its forces from them. During sleep at night, then, man actually expands over the whole solar system". (Matthew 58).

The explanation continues: "What is it that permeates the astral body during sleep when it is outside the physical and etheric bodies? "It is the weaving life of the harmonies of the spheres, forces that can otherwise operate only in the sound-ether. Just as when a violin bow is drawn across the edge of a metal disc strewn with sand the vibrations pulsing through the air also pulsate through the sand and produce the well-known Chladni sound figures, SO DO THE HARMONIES OF THE SPHERES VIBRATE THROUGH THE HUMAN BEING DURING SLEEP AND BRING ORDER AGAIN INTO WHAT HAS BEEN CAST INTO DISORDER DURING THE DAY THROUGH HIS SENSE PERCEPTIONS."

In the lecture, The Descent of the Spirit, Rudolf Steiner characterizes it as follows: "While we are sleeping the starry heavens become our world in exactly the same way that the earth is our world when we are awake. The human soul feels united with the planetary world and we have within us the forces of the movements of the planets immediately after going to sleep and just before waking.

"We carry in us as likeness a little image, a spirit orb, in which the movements of the planets are actually portrayed, and THIS IS INDIVIDUAL IN EACH INDIVIDUAL. Inwardly in a sort of planetary globe in his astral body, he has an imitative experience of all that takes place out in cosmic space among the planets as they move."

Consider the mechanical devices men have invented to 'imitate' the machine recording transactions in the stock exchanges to machines in all parts of the world.

The soul and Ego expands in the second period after going to sleep, into the vastness of the zodiac. In the bewildering vastness of the twelve constellations, "Christ becomes the Guide . . . Simply to have felt what it means for the earthlife that the Christ has passed through the Mystery of Golgotha, this will bring into the astral body . . . . that the Christ becomes the Guide in the zodiac from the time we go to sleep till we wake".

While the Ego and astral body are in the cosmos, the physical body and etheric body are not deserted. Hierarchical Beings must guard the physical body as well as the etheric body. They cannot be abandoned without harm.

The Third Hierarchy (Angels, Archangels and Archai) lives in the same organs in which by day our ego lives, that is, the "pulsing forces of the blood".

During the day, the astral body lives within the breathing organs. There, "the Beings of the next higher hierarchy, Spirits of Form, Spirits of Motion and Spirits of Wisdom, are active during the night."

"Regarding the etheric body, we are not capable, even during the day of doing what must be done. This the beings of the highest hierarchy, Thrones, Cherubim and Seraphim, must fulfill, even when we are awake; they remain permanently in the etheric body". There lie the wonders which are produced by our glands, the seat of the etheric body!

And our physical body! We are completely impotent out of our forces to carry out the magnificent processes which take place within our physical organism. These require the highest powers. "These Powers are none other than those known since primeval times as the Powers of the Highest Trinity, the Powers of the Father, Son and Holy Spirit. The real Trinity dwells in our physical body". (Man As An Image of Spiritual Beings).

CHAPTER IX

Reincarnation and Karma

The word reincarnation, up to very recently, sounded strange to western ears in ordinary everyday conversation. Reincarnation was considered to be a superstition of less developed cultures.

Christians think it is an oriental teaching not taught by Jesus Christ. There are many things which Jesus obviously did not teach and which He must have known since He is a Cosmic Being. For instance, did Jesus teach that the earth is round or that the sun is the center of the solar system? No, He did not. Why not?

The obvious reason is that mankind was not ready at that time for that knowledge to be known generally. The western hemisphere had to be kept from European humanity for a time, although it was known in certain circles.

For one thing, the etheric forces of the west are different from those of Europe. These are the forces affecting living plant species, animal species and even the size and features of men. The major features of the earth, the mountain chains, run north-south in the Americas, whereas the European-Asiatic mountain chains run east-west. The electro-magnetic forces which have their center of intensity at the magnetic pole in North America require a more earth-bound soul to cope with them.

REFERENCES TO REINCARNATION IN THE BIBLE

Jesus did speak of reincarnation when He said of John the Baptist that " 'Elias is come already and they knew him not'. Then the disciples understood that He spake to them of John the Baptist". (Matthew 17:13)

In the annunciation of the angel Gabriel to Zacharias is the statement: "and thy wife Elizabeth shall bear thee a son and thou shalt call his name John. And he shall go before him in the spirit and power of Elias." (Luke 1:13-16) There we read the words "in the spirit and power of Elias", pointing to the spirit, the ego of Elias that was to reincarnate as John, later known as the Baptist.

In the Gospel of St. Mark; "Jesus asked his disciples, saying unto them, Whom do men say that I am?

"And they answered, John the Baptist: but some say Elias; and others, One of the prophets." (Mark 8:27,28 and Matthew 16:13-16).

The questions and answers all point to the fact that reincarnation was widely accepted as reality in Jesus' time and that Jesus confirmed it in the case of John the Baptist.

The fact that Elias was indeed reincarnated as John the Baptist has been reconfirmed by the independent researches of the Akashic Records by Rudolf Steiner. Several other earlier and later incarnations are known of this great leader of humanity.

Most educated people in the Christian world believe that they live only one life on earth and that they then go to heaven or hell, apparently to remain there forever. Others believe that a life after death is only a doubtful possibility. These people therefore feel it is safer to follow externally imposed commandments, just in case there should actually be a life hereafter.

There are also those who do not believe in a life after death. They can then conclude that they are not accountable for their deeds on earth. A callous one can take the attitude that a life of crime is logical. Such activities can take the most varied forms: for instance, getting the means for living by the easiest ways possible such as by peddling drugs, mugging, murder, purse-snatching, burglary, robbing widows and the aged, stealing automobiles, shop-lifting, dependence by the able-bodied on welfare as a way of life, bombing of public and private areas: all these evils will become rampant in a civilization that is living in ignorance of reincarnation and karma.

All sorts of unconscionable acts are perpetrated: reporting events or speaking in such a way as to create hate, suspicion, intolerance, injustice and inflame others to destructive acts. The reader should be able to add many others to this list.

KARMA

Karma is the spiritual law of cause and effect, not as it works from one day to the next day, but as it works from one life on earth to the next life on earth! The laws of karma must become known as a reality of life to the common man if crime, hate-mongering, unconscionable acts of every sort are to be eliminated from the world.

If a man builds the foundation of a house one day and he continues working on it the following days, that would not be called karma. If the house he builds is faulty, collapses and injures or kills the tenants, he may go to jail for his errors.

What results in his next incarnation as the result of the faulty house he had built and what he then has to do to compensate for his errors, is karma. The cause is in one life and the effect extends to the next earth life and beyond.

That is why so many events we see around us do not seem to have 'rhyme or reason'. We cannot see why misfortune strikes one person and misses the one next to him. The unseen cause is in the previous life on earth, and its effect is in the life observed.

That also is the reason for the vast differences in talents we find among children in the same family. It is we who choose our parents, as is shown in the section on the return of the ego from the zodiac and beyond. (pages 66-74).

One might ask, "Why do so many choose such difficult situations in life?".

While in the cosmos in the life between death and rebirth, the judgment of the ego is for overall progress in the coming incarnation, and a difficult earthlife may be the only way for the ego to gain the lessons it requires, even if these involve tragedy, sickness, pain and poverty.

What is reincarnation and what is known about it? What have the higher faculties of perception discovered about reincarnation? What do the Akashic Records reveal about it? Is reincarnation repeated earth lives?

The word 'repeated' gives the impression of living the same life over again. Reincarnation is certainly not that. Reincarnation has a more fundamental purpose than just repeating. It embodies the possibility of advancing, of progress, of a forward evolution of the human individuality, of the development of the ego powers of man to a higher and higher level. It provides opportunities for developing faculties which physical life on this earth alone gives. This 'speck of dust in cosmic space' is exceedingly more important than astronomers and scientists who judge by size alone, attribute to it.

Earth is a unique combination of physical conditions in our solar system. Earth is the only planet in our solar system where physically visible beings such as we humans exist in bodily form. However, discarnate human beings leaving earth-life and also those heading towards earthlife are in all the planetary spheres, as mentioned in the sections dealing with those subjects.

The study of reincarnation must be centered on understanding the laws which can be observed as to how the deeds in one life affect the following lives. The knowledge of reincarnation and karma gives the knower a feeling of responsibility in all of life's relationships. "A man who realizes that the ideas of reincarnation and karma are based on reality . . . knows that men's lives will be shaped according to what they have been in earlier incarnations". (Reincarnation and Karma 86). These ideas will pass over into the souls and hearts of men and arise as moral impulses.

"The incarnating spirit brings its destiny with it from its previous incarnations. Those people with whom the soul was bound up in one life, the soul must meet again in a subsequent one, because the actions which have taken place between them, must have their consequences. The life of the soul is therefore the result of the self-created destiny of the human spirit. The soul is subject to its self-created fate. One calls this fate, created by the man himself, his karma". (Theosophy 54).

At the beginning of life we are drawn to certain others through blood relationship. As a rule "...the human beings with whom we came to be associated involuntarily at the beginning of our life were close to us in a former life, in most cases in the immediately preceding life, in middle life, in the thirties. Then we chose them voluntarily in some way, drawn to them, perhaps by our hearts." (Reincarnation & Karma).

We can therefore look even in this life at around our thirtieth year, for those who influence us greatly, to discover those who could be our parents or other blood relative in our next incarnation.

THE PERIOD OF THE EGO'S RETURN TO EARTH

In the normal course of post Atlantean civilization, the ego returns to earth when the earth has changed sufficiently to provide new conditions for the development of the ego.

One measure of this change of cosmic influences on earth is the position of the sun at the spring equinox. This point makes a circuit of the entire heavens in about 26,000 years, a period known as the Platonic Year. In a twelfth of this period, about 2100 years, the cosmic influences have shifted to another group of Beings of the zodiac.

Man therefore experiences life on earth in a different environment. He also experiences it differently whether he is in a male body or in a female body.

In general, therefore, two incarnations take place in this period of 2100 years, one in a male body and one in a female body. However, there are numerous factors altering this period of about one thousand years between incarnations.

In our epoch of culture, the normal period has speeded up from a thousand years to several hundred years. In addition, the great changes in earth conditions in the last few centuries have been so rapid and far-reaching due to the development of the scientific age that souls are rushing back sooner in order to experience our materialistic civilization.

Individual, national and world destiny all have their influence on the frequency of return to earth life. It can well be that some human souls are at such a low state of development or be so evil in their nature that they are unable to benefit from the usually long sojourn in the cosmos. They therefore return sooner than usual. "Ahrimanic beings also exert an influence to shorten the period from the cosmos back to earth by as much as two hundred years."

Certain special egos return to earth again and again in a male or female body, up to a maximum of seven times, without the compensating experience in a different sex.

Since it is a need for the ego's development to experience life in the widest variety of conditions and to gain diversified talents, the egos usually incarnate in a different race when they return.

We have all developed and lived through the various epochs of history as well as through the planetary evolutions of earth. So it is we of whom we study when we delve into cosmic and human history.

THE VARIATION AMONG INDIVIDUALS

How the egos utilize each life or how they fail to make use of their opportunities can be seen in every classroom. The strongest and most conscious egos can make great advances in one life, while others seem to fall backward instead of making progress. Over a number of earth lives, the differences among individuals can be tremendous.

The uniqueness of each human being stems from the varied progress made during many earthly lives. Some rise to great heights, while others, because of love of ease, laziness, or self-imposed lower states of consciousness make slower progress. "It is known that Mozart when a boy, could write out from memory a long musical work of art after hearing it only once".

The geneticists' idea that genes account for

all the variations of talents among human beings falls far short of the Reality.

Knowledge of the laws of reincarnation and karma would show men that:

they return to earth and meet those with whom they had lived; those they had wronged, as well as those they had loved;

they return in different nations and races;

they may have to incarnate later in a race they hated and despised in this life, in order to experience the hatred they espoused;

the karmic consequence of a one-sided male life is a female incarnation in the next life;

the karmic consequence of a one-sided female life is a male incarnation in the next life;

each race and people develop certain talents and qualities to a higher level than other races, so that mankind as a whole, and as individuals also, become heir potentially to the highest level developed;

each epoch of culture develops specific human capacities and that man must not permit those opportunities to slip by because of his drowsiness and his sleepy-headedness.

If men lived in that knowledge, racial hatred may be turned to gratefulness for the contribution of other races, and racial appreciation can supplant racial prejudice.

Then all the other problems of existence would fall into order and men would take a more responsible attitude towards their acts, as well as toward their thoughts and feelings of the world and for the world.

The seeker of knowledge of how karma works from life to life can gain invaluable insight by a study of the six volumes entitled KARMIC RELATIONSHIPS, Volume 1 to Volume 6, containing a total of sixty-six lectures on the previous life or lives of many historical characters. One should not quote from these lectures and try to sensationalize the information contained in them, but must use the examples given to gain deeper understanding of how one's own life was determined by his past life and how his present life determines his future life on earth.

The student can also come to see that it is most important that he does not allow his egotism to cause him to imagine that he was someone important in a previous life, and attempt to live as he had been. He must realize that his own life situation and condition was determined by a wisdom far higher than his own day-consciousness, to be the best situation he needs in order to achieve the development in soul and spirit which he requires in this moment of his individual evolution.

He comes to realize he himself had wilfully chosen the very conditions he finds in life. The smaller and larger groups to which he belongs are also determined by his relationship with them in the past lives. Those we meet in middle life, we meet again in childhood, and in another incarnation, late in life.

Many indications are given in various lectures, such as, "A person who tells many untruths or is inclined to take things superficially will be a thoughtless character in the next or a later incarnation". (Reincarnation & Karma 93).

The FOUR MYSTERY PLAYS by Rudolf Steiner portray in dramatic form the way karma and reincarnation works. These plays are THE PORTAL OF INITIATION, THE SOUL'S PROBATION, THE GUARDIAN OF THE THRESHOLD and THE SOUL'S AWAKENING.

A group of twentieth century people tied together by karmic relationships are traced back to the fourteenth century, then a glimpse in Hibernian times and then far back to about 2,000 B.C. in the temples of Egypt. The soul world and beings are depicted as the Realities which they are to the clairvoyant faculties of an Initiate. How they work to influence and seduce men in the case of Lucifer and Ahriman, never tiring in their efforts, will probably be a shock to many who think they are so worldly wise.

The leading figures are seen again and again, guiding, teaching, assisting; certain others, opposing in one life, but uniting in a later life those whom they had opposed! The ramifications of karma are complex and these plays require and deserve long study.

They are said, by their author, to contain

the whole of the science of the spirit, that is, of anthropoposophy.

REINCARNATION As A Phenomenon of Metamorphosis by Dr. Guenther Wachsmuth is a comprehensive survey of the many aspects of reincarnation revealed in the thousands of lectures of Rudolf Steiner. Dr. Wachsmuth, one of the leading pupils and assistants of Dr. Steiner, made special studies on the etheric formative forces and their metamorphosis. His researches add to the value of the book.

A few of the points considered, selected here and there from that book include mainly Rudolf Steiner's findings in his spiritual researches.

METAMORPHOSIS

Characteristics in people can be traced from one life to the following incarnations. In any one life, the astral body is most easily affected. The process of metamorphosis enters into a deeper level in the next incarnation, that is, the effect will be in the etheric body in the second incarnation. In the third incarnation, the continued effect, if not changed for the better, enters into the physical body, and may result in deformed organs.

As an example, superficiality in the course of one life becomes an inner inclination to untruthfulness in the next life and a disturbance in the processes of organ-formation in the third earthly life. A fickle life showing no love and devotion manifests as a tendency to lying in the next life followed by wrongly developed organs and a tendency to physical illness in the third incarnation.

Not all illnesses are traced back to such causes. It is erroneous to trace everything in life back to this or that guilt of the past. Just as frequently we shall have to think that whatever is seen in life has no relationship to the past, but rather will be the cause for a karmic compensation in the future. Overcoming or experiencing undeserved difficulties in life also leads to strengthening of forces for the future.

An example was cited where a person, because of an undeveloped brain, was condemned to lead a life of idiocy. In the life after death "he was able to work over in himself all the depressing experiences of such a life, the being thrust here and there, the lovelessness of human beings. He was reborn as the very genius of charity".

Another situation may arise where a person who had many bad experiences in a previous life, may, just upon entering into the physical body when he has a prevision of the coming earthlife, draw back and not enter the body completely and the union with his different members in not perfectly established. This also results in idiots in this life, but it is not always the cause.

Other causes of a weak total organization include a lack of interest in the cosmic environment in the previous existence. Some people are so involved with themselves, they never look up at the sky or take any interest in what happens in the heavens.

Those persons who enjoy exuberant health have possessed in a previous earthlife a vivid interest in the visible world. Also, a sound physical organization results in one who had good inclinations in previous lives.

The disposition which leads a person to infectious diseases rests upon a highly developed egotistic acquisitive sense in a preceding life. The egotistic urge to acquisition and gain is a characteristic of the ether-body which appears in the next life as a disposition toward infectious diseases. It is also the case that the opposite extreme may appear, that those of excessively weak self-feeling in the previous existence will have a predisposition to infectious diseases. The forces gained by undergoing infectious illnesses bring about the balance which the conditions in the previous life must have.

****

"Persons who have lost their lives in natural disasters such as earthquakes, for example, became spiritual persons in the next incarnation. They have advanced so far that this single blow was needed to show them the transitoriness of the earthly. This had so strong an after-effect in the existence in the cosmic-spiritual world that they learned as the fruit for the next life that the material is the evanescent, the spiritual is the dominant".

****

Those who die an early death from disease or accident have unused forces which would have carried them to a normal life-span. Spiritual research finds that these unused forces are 'preserved as a residue' and these forces are added to the forces of will and feeling of the person and they fortify him in his next earthlife. He would therefore be stronger and able to accomplish more in that later life.

This is true only where the wisdom-filled karmic relationships determine such earlier deaths. "It is not so in the case of suicides. Suicides form judgments based on only a small fraction of the complex laws, great interrelationships and the rhythms involved in human incarnation, as to what age in history, the groups of souls with whom we must incarnate, and ramifications too complex for the earth-bound human to conceive.

Suicide is not only the destruction of the physical body, brought about out of an earthly, subdued consciousness, but a disturbance contrary to all spiritual laws of evolution, if not, perhaps the destruction of the finest inner structure of the human being." By his act, the suicide isolates himself from the physical world violently, and also isolates himself from the spiritual world as in a cell antipathetic to all existence.

"The loneliness of the soul of the suicide exceeds anything imaginable on the physical plane," and the spiritual investigator says 'if the suicide knew its effect upon his future existence, and were conscious that he does not escape in any way from his destiny, but only makes it more difficult, he would never perform this violent interference in the laws of his future life . . . . Suicide is never a solution in accordance with the spirit, but an intensification of the difficulties which the human being must overcome upon the ascending evolutionary path".

****

The thinking of man determines the future of the human race. "What humanity thinks today, that will it be in the future. A humanity that thinks materialistically will bring forth terrible beings in the future; and a humanity that thinks spiritual thoughts has such a transforming influence upon the organism of the future that beautiful human bodies will arise. All materialistically thinking souls work for the bringing forth of evil race-formations, and all that is spiritually wrought causes the bringing forth of a good race".

FREEDOM

Does the idea of freedom and karma conflict? "Man is free at every moment to enter new items in the karmic book of life. Let no one ever believe, therefore, that an irrevocable law of destiny is fixed for life. Freedom is not affected by the law of karma. For this reason, we must think just as much about the future in connection with the law of karma as about the past. We bear within us the effects of past acts and we are the slaves of the past, BUT THE MASTERS OF THE FUTURE".

How much we become masters of the future depends on how conscious we are of the workings of the law of karma and reincarnation, and on what efforts we make to carry forward our own evolution.

The first step is in thinking in conformity with the REALITY of the nature of man, the cosmos and the Spiritual Beings. "It is necessary to know that it is immoral to hold a false theory, especially in those fields of knowledge related to the being of man and the phenomena of life.

"If men should shut themselves off further from the knowledge and forces of the supersensible world, it would gradually come about that when men returned again to earthly incarnations, they would not possess sufficient life-giving forces, such as they need. A decadence, a withering away of the physical humanity would occur if the spiritual forces should not be taken up."

We see already signs of this in the sclerotic condition among growing numbers of the young. This will spread so less and less human beings will be able to bear children.

Humanity stands at the threshold of the spiritual world.

His destiny is to cross the threshold consciously while he is in the physical world.

And he can, if he will.

BIBLIOGRAPHY

The following books are by Rudolf Steiner, Ph.D. (1861-1925)

Theosophy
Knowledge of the Higher Worlds and Its Attainment
Occult Science--An Outline
Cosmic Memory
Occult Research Into Life Between Death And A New Birth
The Soul's Progress Through Repeated Earth Lives
Man's Life On Earth And In The Spiritual Worlds
Anthroposophic Ethics
The Gospel of St. Matthew
The Gospel of St. Mark
The Gospel of St. Luke
The Gospel of St. John
The Gospel of St. John in Relation to Other Gospels
The Pre-Earthly Deeds of Christ
Secrets of the Bible Story of Creation
The True Nature of the Second Coming
From Jesus to Christ
The Fifth Gospel
Spiritual Guidance of Man and of Mankind
Reincarnation and Karma
Karma
The Karma of Vocations
Four Mystery Plays
Vol. I The Portal of Initiation
The Soul's Probation
Vol. II The Guardian of the Threshold
The Soul's Awakening
Excursus on the Gospel According to St. Mark

The Bible and Wisdom
The Apocalypse of St. John
The Influence of Spiritual Beings Upon Man
Descent of the Spirit
Man As An Image of Spiritual Beings
Lucifer and Ahriman
Vision of Christ's Advent in the Etheric
Manifestations of Karma
Karmic Relationships Vol. 1 to Vol. 6
Deeds of Christ And The Opposing Powers

Sources of Rudolf Steiner Books

Anthroposophic Press, Inc. 258 Hungry Hollow Road, Spring Valley, New York 10977

Multimedia Corp. 100 S. Western Hwy., Blauvelt, N.Y. 10913

St. George Book Service, P.O. Box 225, Spring Valley, N.Y. 10977

Rudolf Steiner Library, Anthroposophical Society in America, 211 Madison Avenue, New York, N.Y. 10016

Steiner Book Centre, Inc., 151 Carisbrooke Crescent, North Vancouver, B.C. Canada

Bucherverkauf am Goetheanum, Dornach CH 4143 Switzerland

Rudolf Steiner Book Shop, 35 Park Road, London NW1 6XT, England

Rudolf Steiner Book Shop, 38 Museum St., London WC1 England

Rudolf Steiner Book Shop, 722 Darling St., Rozelle, NSW 2037, Australia

References by other authors

Reincarnation by Guenther Wachsmuth, Ph.D.
Meditation--An Introduction, Ernst Katz, Ph.D.
Professor of Physics, University of Michigan
What Difference Did the Deed of Christ Make, Beredene Jocelyn
(These last two pamphlets are obtainable from Rudolf Steiner Information Center
211 Madison Ave., New York, N.Y. 10016

Order blank:

- -